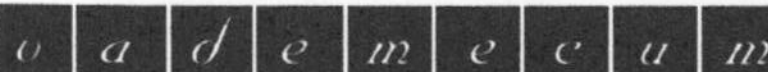

Chemotherapy Regimens and Cancer Care

Alan D. Langerak, M.D.
Cancer Care Associates
Tulsa, Oklahoma

Luke P. Dreisbach, M.D.
The University of Arizona
Tucson, Arizona

CRC Press is an imprint of the
Taylor & Francis Group, an **informa** business

VADEMECUM

Chemotherapy Regimens and Cancer Care

First published 2001 by Landes Bioscience

Published 2018 by CRC Press
Taylor & Francis Group
6000 Broken Sound Parkway NW, Suite 300
Boca Raton, FL 33487-2742

ISBN-13: 978-1-57059-670-4 (pbk)

Visit the Taylor & Francis Web site at
http://www.taylorandfrancis.com

and the CRC Press Web site at
http://www.crcpress.com

Library of Congress Cataloging-in-Publication Data

Langerak, Alan D.
Chemotherapy regimens and cancer care / Alan D. Langerak, Luke P. Dreisbach.
p.; cm. -- (Vademecum)
Includes index.
1. Antineoplastic agents--Handbooks, manuals, etc. 2. Cancer--Chemo therapy--Handbooks, manuals, etc. I. Dreisbach, Luke P. II. Title. III. Series.
[DNLM: 1. Antineoplastic Agents, Combined--administration & dos age--Handbooks. 2. Antineoplastic Agents, Combined--therapeutic use --Handbooks. 3. Neoplasms--drug therapy--Handbooks. QV 39 L276c 2001]
RC271.C5 L287 2001
616.99'4061--dc21 2001029156

Dedication

To Sue and Kelly:
for all of their love and support.

Contents

Preface

Chemotherapy Regimens and Cancer Care is a book that is intended for practical use by Hematology/Oncology attendings, fellows, residents, pharmacists, and nurses. It is a concise, thorough, and portable reference guide to the multitude of complex chemotherapy protocols and other frequently utilized medications in the field of Hematology/Oncology.

This book was designed to be different from other "chemotherapy handbooks" in several ways. First, the book summarizes the most commonly used oncology regimens and gives practical guidelines for the supportive care required for optimal administration of these regimens. The regimens include not only a reference, but also recommended antiemetics and helpful reminders about unique toxicities of the various chemotherapeutic agents. The protocols are arranged in a disease-based manner for easy accessibility. Within each section, combination regimens are listed first, in alphabetical order, followed by single agents with activity in that disease. We have included many of the more commonly used chemotherapy protocols, as well as some newer, promising protocols; however, the listing of regimens is not all-inclusive or meant to direct therapy.

Second, the book includes sections on pain control, antibiotic use in neutropenic fever, antiemetic guidelines, and supportive care medications. The book continues with a section on drugs used in commonly encountered problems in hematology, and concludes with a listing of the wholesale costs of most chemotherapy agents. The rapid growth of chemotherapeutic options will make frequent updates of this handbook essential for the future state-of-the-art care of our patients.

During our educational journey into the ever-evolving fields of hematology and oncology, we noticed that there was no updated, well-organized guide, as described above. We envision this book to be utilized on a regular basis by all those involved in the day-to-day care of patients with cancer. We sincerely hope our efforts in preparing this handbook improve the care of those suffering from cancer. This handbook has already paid dividends in assisting us to better care for our patients, and we hope that you, the reader, will also be pleased.

Alan D. Langerak, M.D.
Luke P. Dreisbach, M.D.

Acknowledgments

We wish to thank Harinder Garewal, M.D., Ph.D. of the Arizona Cancer Center and Tucson V.A. Medical Center for all of his guidance and support.

We would also like to express our appreciation to the faculty of the Arizona Cancer Center at the University of Arizona in Tucson for the knowledge they imparted to us which allowed us to share the information in this book with our readers.

A.L. and L.D.

Chapter 1
Brain Cancer

Chemotherapy Regimens and Cancer Care, by Alan D. Langerak and Luke P. Dreisbach.

Brain Cancer

Agent	Dosage
Carmustine (BCNU)	BCNU 80 mg/M² IV days 1-3 REF: Walker et al. J Neurosurg 1978; 49:333-343 PREMEDICATIONS 1. Kytril 1 mg PO/IV 30 minutes before and 12 hours after chemotherapy on days 1-3 2. Dexamethasone 10 mg IV before chemotherapy on days 1-3 Give non-cisplatin delayed emesis prophylaxis Repeat every 6-8 weeks Carmustine—maximum total dose is 1440 mg/M²; causes delayed myelosuppression
PCV (standard dose) procarbazine/ lomustine (CCNU)/ vincristine	Procarbazine 60 mg/M² PO days 8-21 CCNU 110 mg/M² PO day 1 Vincristine 1.4 mg/M² IV days 8,29 –this regimen is started within 14 days of completion of XRT REF: Levin et al. Int J Radiat Oncol Biol Phys 1990; 18:321-324 PREMEDICATIONS 1. Kytril 1 mg PO/IV 30 minutes before and 12 hours after chemotherapy on day 1 2. Dexamethasone 20 mg IV before chemotherapy on day 1 Repeat every 42 days until progression or a maximum of one year Lomustine—delayed myelosuppression Vincristine—vesicant–avoid extravasation; cumulative neurotoxicity—may produce severe constipation –maximum 2 mg per administration
I-PCV (intensified) procarbazine/ lomustine (CCNU)/ vincristine	CCNU 130 mg/M² PO day 1 Procarbazine 75 mg/M² PO days 8-21 Vincristine 1.4 mg/M² IV days 8,29 –no dose maximum for Vincristine REF: Cairncross et al. J Clin Oncol 1994; 12:2013-2021

Continued

Agent	Dosage
	PREMEDICATIONS 1. Kytril 1 mg PO/IV 30 minutes before and 12 hours after chemotherapy on day 1 2. Dexamethasone 20 mg IV before chemotherapy on day 1 Repeat every 6 weeks Lomustine—delayed myelosuppression Vincristine—vesicant–avoid extravasation; cumulative neurotoxicity—may produce severe constipation; –maximum 2 mg per administration
Temozolomide	Temozolomide 150-200 mg/M^2 PO days 1-5 REF: Yung et al. J Clin Oncol 1999; 17:2762-2771 Repeat every 28 days Temozolomide—start at 150 mg/M^2 and advance dose up to 200 mg/M^2 as tolerated, based on myelosuppression (adjust dose per package insert); taken for a maximum of 2 years, or until disease progression
Thalidomide	for high-grade gliomas Thalidomide 800 mg PO QHS daily –dose advanced 200 mg every 2 weeks as tolerated to maximum of 1200 mg daily REF: Fine et al. J Clin Oncol 2000; 18:708-715 Thalidomide—providers and pharmacies must be registered with the S.T.E.P.S. program; can cause significant somnolence

Chapter 2
Breast Cancer

Breast Cancer

Agent	Dosage
AC doxorubicin/ cyclophos-phamide	Doxorubicin 60 mg/M^2 IV day 1 Cyclophosphamide 600 mg/M^2 IV day 1 REF: Fisher et al. J Clin Oncol 1990; 8:1483-1496 PREMEDICATIONS 1. Kytril 1 mg PO/IV 30 minutes before and 12 hours after chemotherapy 2. Dexamethasone 20 mg IV 30 minutes before chemo-therapy Repeat every 21 days Doxorubicin—monitor cumulative dose for cardiac toxicity (not to exceed 550 mg/M^2 or 450 mg/M^2 with prior chest radio-therapy); vesicant—avoid extravasation; use 50% for bilirubin 1.5-3.0; use 25% for bilirubin > 3.0
A → CMF doxorubicin followed by cyclophospha-mide/ methotrexate/ fluorouracil (5-FU)	Doxorubicin 75 mg/M^2 IV day 1 –given every 21 days for 4 cycles Cyclophosphamide 600 mg/M^2 IV day 1 Methotrexate 40 mg/M^2 IV day 1 5-FU 600 mg/M^2 IV day 1 –the CMF portion of this regimen is given every 21 days for 8 cycles REF: Bonadonna et al. JAMA 1995; 273:542-547 PREMEDICATIONS 1. Kytril 1 mg PO/IV 30 minutes before and 12 hours after chemotherapy 2. Dexamethasone 20 mg IV 30 minutes before chemo-therapy Doxorubicin–monitor cumulative dose for cardiac toxicity (not to exceed 550 mg/M^2 or 450 mg/M^2 with prior chest radiotherapy); vesicant—avoid extravasation; use 50% for bilirubin 1.5-3.0; use 25% for bilirubin > 3.0 Methotrexate—use 75% dose for CrCl < 50; 50% dose if CrCl < 25; do not give if patient has an effusion ("reservoir effect")

Agent	Dosage
AC → T doxorubicin/ cyclophosphamide followed by paclitaxel	Doxorubicin 60 mg/M² IV day 1 Cyclophosphamide 600 mg/M² IV day 1 –above combination is given every 3 weeks for 4 cycles, followed by Paclitaxel 175 mg/M² IV day 1 –every 3 weeks for 4 cycles REF: Henderson et al. Proc Amer Soc Clin Oncol 1998; 390A PREMEDICATIONS 1. Kytril 1 mg PO/IV 30 minutes before and 12 hours after chemotherapy (for AC) 2. Dexamethasone 20 mg IV 30 minutes before chemotherapy 3. Cimetidine 300 mg IV 30 minutes before paclitaxel 4. Diphenhydramine 25-50 mg IV 30 minutes before paclitaxel 5. Compazine 10 mg PO/IV 30 minutes before paclitaxel OTHER MEDICATIONS 1. Dexamethasone 4 mg PO BID for 6 doses after paclitaxel (for myalgias) Doxorubicin–monitor cumulative dose for cardiac toxicity (not to exceed 550 mg/M² or 450 mg/M² with prior chest radiotherapy); vesicant—avoid extravasation; use 50% for bilirubin 1.5-3.0; use 25% for bilirubin > 3.0
ATC doxorubicin/ paclitaxel/ cyclophosphamide (sequential)	Doxorubicin 90 mg/M² IV days 1, 15,29 Paclitaxel 250 mg/M² CIV (X 24 h) days 43, 57,71 Cyclophosphamide 3000 mg/M² IV days 85, 99,113 REF: Hudis et al. J Clin Oncol 1999; 17:93-100 PREMEDICATIONS 1. Kytril 1 mg PO/IV 30 minutes before and 12 hours after chemotherapy on days 1, 15, 29, 85, 99, and 113 2. Dexamethasone 20 mg IV 30 minutes before chemotherapy on days 1, 15, 29, 85, 99, and 113 3. Dexamethasone 20 mg IV 30 minutes before chemotherapy on days 43, 57, and 71 OR Dexamethasone 20 mg PO 6 hours and 12 hours prior to chemotherapy on days 43, 57, and 71

Continued

Agent	Dosage
	4. Cimetidine 300 mg IV 30 minutes before chemotherapy on days 43, 57, and 71 5. Diphenhydramine 25-50 mg IV 30 minutes before chemotherapy on days 43, 57, and 71 6. Compazine 10 mg PO/IV 30 minutes before chemotherapy on days 43, 57, and 71 OTHER MEDICATIONS 1. Dexamethasone 4 mg PO BID for 6 doses after paclitaxel (for myalgias) 2. G-CSF through entire course of chemo (5 mcg/kg days 3-10 of each 14 day course) 3. Give non-cisplatin delayed emesis prophylaxis Doxorubicin–monitor cumulative dose for cardiac toxicity (not to exceed 550 mg/M² or 450 mg/M² with prior chest radiotherapy); vesicant—avoid extravasation; use 50% for bilirubin 1.5-3.0; use 25% for bilirubin > 3.0
CAF/IV cyclophos-phamide/ doxorbucin/ 5-FU	Cyclophosphamide 500 mg/M² IV day 1 Doxorubicin 50 mg/M² IV day 1 5-FU 500 mg/M² IV day 1 REF: Smalley et al. Cancer 1977; 40:625-632 PREMEDICATIONS 1. Kytril 1 mg PO/IV 30 minutes before and 12 hours after chemotherapy 2. Dexamethasone 20 mg IV 30 minutes before chemo-therapy Repeat every 21 days Doxorubicin–monitor cumulative dose for cardiac toxicity (not to exceed 550 mg/M² or 450 mg/M² with prior chest radiotherapy); vesicant—avoid extravasation; use 50% for bilirubin 1.5-3.0; use 25% for bilirubin > 3.0
CEF cyclophospha-mide/ epirubucin/ 5-FU	Cyclophosphamide 75 mg/M² PO days 1-14 Epirubicin 60 mg/M² IV days 1, 8 5-FU 500 mg/M² IV days 1, 8 REF: Levine et al. J Clin Oncol 1998; 16: 2651-2658 PREMEDICATIONS 1. Kytril 1 mg PO/IV 30 minutes before and 12 hours after chemotherapy on day 1 2. Dexamethasone 20 mg IV 30 minutes before chemo-therapy on day 1

Continued

<table>
<tr><th>Agent</th><th>Dosage</th></tr>
<tr><td></td><td>OTHER MEDICATIONS
1. Trimethoprim-sulfamethoxazole DS 2 tabs PO BID for duration of chemotherapy

Repeat every 28 days for 6 cycles

Epirubicin–monitor cumulative dose for cardiac toxicity (not to exceed 1000 mg/M²); vesicant—avoid extravasation</td></tr>
<tr><td>CMF/IV cyclophos-phamide/ methotrexate/ 5-FU—21 day</td><td>Cyclophosphamide 600 mg/M² IV day 1
Methotrexate 40 mg/M² IV day 1
5-FU 600 mg/M² IV day 1

REF: Hainsworth et al. Cancer 1997; 79:740-748

PREMEDICATIONS
1. Kytril 1 mg PO/IV 30 minutes before and 12 hours after chemotherapy
2. Dexamethasone 20 mg IV 30 minutes before chemo-therapy

Repeat every 21 days

Methotrexate–use 75% dose for CrCl < 50; 50% dose if CrCl < 25; do not give if patient has an effusion ("reservoir effect")</td></tr>
<tr><td>CMF/IV cyclophos-phamide/ methotrexate/ 5-FU—28 day</td><td>Cyclophosphamide 600 mg/M² IV days 1, 8
Methotrexate 40 mg/M² IV days 1, 8
5-FU 600 mg/M² IV days 1, 8

REF: Harper-Wynne et al. Br J Cancer 1999; 81:316-322

PREMEDICATIONS
1. Kytril 1 mg PO/IV 30 minutes before and 12 hours after chemotherapy on days 1 and 8
2. Dexamethasone 20 mg IV 30 minutes before chemo-therapy on days 1 and 8

Repeat every 28 days

Methotrexate–use 75% dose for CrCl < 50; 50% dose if CrCl < 25; do not give if patient has an effusion ("reservoir effect")</td></tr>
<tr><td>CMF/PO cyclophos-phamide/ methotrexate/ 5-FU</td><td>(Bonadonna regimen)
Cyclophosphamide 100 mg/M² PO days 1-14
Methotrexate 30-40 mg/M² IV days 1, 8
5-FU 400-600 mg/M² IV days 1, 8
–use lower doses listed for age > 65

REF: Bonadonna et al. NEJM 1976; 294:405-410</td></tr>
</table>

Continued

Agent	Dosage			
	PREMEDICATIONS 1. Compazine 10 mg PO/IV 30 minutes before chemotherapy on days 1 and 8 Repeat every 28 days Methotrexate–use 75% dose for CrCl < 50; 50% dose if CrCl < 25; do not give if patient has an effusion ("reservoir effect")			
FEC fluorouracil (5-FU)/ epirubicin/ cyclophosphamide	5-FU	500 mg/M²	IV	day 1
	Epirubicin	60 mg/M²	IV	day 1
	Cyclophosphamide	500 mg/M²	IV	day 1
	REF: Blomqvist et al. J Clin Oncol 1993; 11:467-473 PREMEDICATIONS 1. Kytril 1 mg PO/IV 30 minutes before and 12 hours after chemotherapy 2. Dexamethasone 20 mg IV 30 minutes before chemotherapy Repeat every 28 days Epirubicin–monitor cumulative dose for cardiac toxicity (not to exceed 1000 mg/M²); vesicant – avoid extravasation			
MV mitomycin C/ vinblastine	Mitomycin C	12 mg/M²	IV	day 1
	Vinblastine	6 mg/M²	IV	days 1,22
	REF: Nabholtz et al. J Clin Oncol 1999; 17:1413-1424 PREMEDICATIONS 1. Kytril 1 mg PO/IV 30 minutes before and 12 hours after chemotherapy on day 1 2. Dexamethasone 20 mg IV 30 minutes before chemotherapy on day 1 3. Compazine 10 mg PO/IV before chemotherapy on day 22 Repeat every 42 days Vinblastine–use 50% of dose for bilirubin > 3.0; vesicant–avoid extravasation; watch for neurotoxicity Mitomycin C–myelosuppression occurs late (approximately 4 weeks); limit cumulative dose to 50 mg/M² (vascular toxicity)			
TA docetaxel/ doxorubicin	Docetaxel	75 mg/M²	IV(over 1h)	day 1
	Doxorubicin	50 mg/M²	IV	day 1
	REF: Dieras et al. Oncology 1997; 11:(8 Suppl 8):31-33			

Continued

Brain Cancer | Breast Cancer | Carcinoma of Unknown Primary | Endocrine Cancer | Gastrointestinal Cancer | Genitourinary Cancer | Gynecologic Cancer | Head and Neck Cancer | Hematologic Malignancies

Agent	Dosage
	OR Docetaxel 60 mg/M² IV(over 1h) day 1 Doxorubicin 60 mg/M² IV day 1 REF: Sparano et al. J Clin Oncol 2000; 18:2369-2377 PREMEDICATIONS 1. Dexamethasone 20 mg IV 30 minutes before chemotherapy 2. Cimetidine 300 mg IV 30 minutes before chemotherapy 3. Diphenhydramine 25-50 mg IV 30 minutes before chemotherapy 4. Kytril 1 mg PO/IV 30 minutes before and 12 hours after chemotherapy OTHER MEDICATIONS 1. Dexamethasone 8 mg PO BID for 8 doses—start day prior to chemo (decreases lower extremity edema) Repeat every 21 days Doxorubicin–monitor cumulative dose for cardiac toxicity (not to exceed 550 mg/M² or 450 mg/M² with prior chest radiotherapy); vesicant—avoid extravasation; use 50% for bilirubin 1.5-3.0; use 25% for bilirubin > 3.0
2M mitoxantrone/ methotrexate	Mitoxantrone 6.5 mg/M² IV day 1 Methotrexate 30 mg/M² IV day 1 REF: Stein et al. Eur J Cancer 1992; 28A:1963-1965 PREMEDICATIONS 1. Compazine 10 mg PO/IV 30 minutes before chemotherapy Repeat every 21 days Methotrexate–use 75% dose for CrCl < 50; 50% dose if CrCl < 25; do not give if patient has an effusion ("reservoir effect") Mitoxantrone–watch cumulative dose—do not exceed 140 mg/M²; possible cardiac toxicity
VATH vinblastine/ doxorubicin/ thiotepa/fluoxymesterone	Vinblastine 4.5 mg/M² IV day 1 Doxorubicin 45 mg/M² IV day 1 Thiotepa 12 mg/M² IV day 1 Fluoxymesterone 30 mg PO days 1-21 REF: Hart et al. Cancer 1981; 48:1522-1527

Continued

Agent	Dosage
	PREMEDICATIONS 1. Kytril 1 mg PO/IV 30 minutes before and 12 hours after chemotherapy 2. Dexamethasone 20 mg IV 30 minutes before chemotherapy Repeat every 21 days Doxorubicin–monitor cumulative dose for cardiac toxicity (not to exceed 550 mg/M² or 450 mg/M² with prior chest radiotherapy); vesicant – avoid extravasation; use 50% for bilirubin 1.5-3.0; use 25% for bilirubin > 3.0 Vinblastine–use 50% of dose for bilirubin > 3.0; vesicant–avoid extravasation; watch for neurotoxicity
Capecitabine	Capecitabine 2510 mg/M²/d divided BID PO days 1-14 REF: Blum et al. J Clin Oncol 1999; 17:485-493 PREMEDICATIONS 1. Compazine 10 mg PO before chemotherapy prn Repeat every 21 days
Docetaxel	Docetaxel 100 mg/M² IV(over 1h) day 1 REF: Nabholtz et al. J Clin Oncol 1999; 17:1413-1424 Repeat every 21 days OR Docetaxel 40 mg/M² IV(over 1 h) day 1 REF: Burstein et al. J Clin Oncol 2000; 18:1212-1219 Repeat every 7 days for 6 weeks, followed by a 2 week rest PREMEDICATIONS 1. Dexamethasone 20 mg IV 30 minutes before chemotherapy 2. Cimetidine 300 mg IV 30 minutes before chemotherapy 3. Diphenhydramine 25-50 mg IV 30 minutes before chemotherapy 4. Compazine 10 mg PO/IV 30 minutes before chemotherapy OTHER MEDICATIONS 1. Dexamethasone 8 mg PO BID for 8 doses—start day prior to chemo (decreases lower extremity edema)

Agent	Dosage
Gemcitabine	Gemcitabine 1000 mg/M² IV days 1,8, 15 REF: Possinger et al. Anticancer Drugs 1999; 10:155-162 PREMEDICATIONS 1. Compazine 10 mg PO/IV 30 minutes before chemotherapy Repeat every 28 days
Liposomal Doxorubicin (Doxil)	Doxil 45-60 mg/M² IV day 1 REF: Ranson et al. J Clin Oncol 1997; 15:3185-3191 PREMEDICATIONS 1. Kytril 1 mg PO/IV 30 minutes before and 12 hours after chemotherapy 2. Dexamethasone 20 mg IV 30 minutes before chemotherapy Repeat every 21-28 days Doxorubicin–monitor cumulative dose for cardiac toxicity (not to exceed 550 mg/M² or 450 mg/M² with prior chest radiotherapy); vesicant—avoid extravasation; use 50% for bilirubin 1.5-3.0; use 25% for bilirubin > 3.0
Paclitaxel	Paclitaxel 175 mg/M² IV (over 3 h) day 1 REF: Nabholtz et al. J Clin Oncol 1996; 14:1858-1867 PREMEDICATIONS 1. Dexamethasone 20 mg IV 30 minutes before chemotherapy OR Dexamethasone 20 mg PO 6 and 12 hours prior to chemotherapy 2. Cimetidine 300 mg IV 30 minutes before chemotherapy 3. Diphenhydramine 25-50 mg IV 30 minutes before chemotherapy 4. Compazine 10 mg PO/IV 30 minutes before chemotherapy OTHER MEDICATIONS 1. Dexamethasone 4 mg PO BID for 6 doses after (for myalgias) Repeat every 21 days
Pamidronate	Pamidronate 90 mg IV day 1 REF: Theriault et al. J Clin Oncol 1999; 17:846-854 Repeat every 28 days

Agent	Dosage			
Trastuzumab (Herceptin)	Herceptin	4 mg/kg	IV (over 90 min)	day 1
	Herceptin	2 mg/kg	IV (over 30 min)	weekly thereafter
	–initial infusion is over 90 min; if well-tolerated, subsequent doses are given over 30 min. REF: Cobleigh et al. J Clin Oncol 1999; 17:2639-2648 PREMEDICATIONS 1. Benadryl 25-50 mg PO/IV 30 minutes before Herceptin 2. Tylenol 650 mg PO 30 minutes before Herceptin Repeat every 7 days Trastuzumab–monitor for cardiotoxicity; increases with concurrent Doxorubicin			
Vinorelbine	Vinorelbine	30 mg/M²	IV (over 20 min)	weekly
	REF: Fumoleau et al. J Clin Oncol 1993; 11:1245-1252 PREMEDICATIONS 1. Compazine 10 mg PO/IV 30 minutes before chemotherapy Repeat every 7 days Vinorelbine–vesicant; avoid extravasation; can cause peripheral neuropathy			
Hormonal Agents	Tamoxifen (Nolvadex)	20 mg	PO	QD
	Anastrazole (Arimidex)	1 mg	PO	QD
	Exemestane (Aromasin)	25 mg	PO	QD
	Toremifene (Fareston)	60 mg	PO	QD
	Letrozole (Femara)	2.5 mg	PO	QD
	Megestrol (Megace)	40 mg	PO	QID

Chapter 3
Carcinoma of Unknown Primary

Carcinoma of Unknown Primary

<table>
<tr><th>Agent</th><th colspan="4">Dosage</th></tr>
<tr><td rowspan="3">EP (PE)
cisplatin/
etoposide
(VP-16)</td><td>VP-16</td><td>80-120 mg/M²</td><td>IV</td><td>days 1-3</td></tr>
<tr><td>Cisplatin</td><td>60-100 mg/M²</td><td>IV</td><td>day 1</td></tr>
<tr><td colspan="4">REF: There are multiple variants of this regimen

PREMEDICATIONS
1. Kytril 1 mg PO/IV 30 minutes before and 12 hours after cisplatin
2. Dexamethasone 20 mg IV 30 minutes before cisplatin
3. Compazine 10 mg PO/IV 30 minutes before etoposide

OTHER MEDICATIONS
1. Give cisplatin delayed emesis prophylaxis

Repeat every 21-28 days

Cisplatin—vigorous hydration is required; can be nephrotoxic and ototoxic; can cause peripheral neuropathy; hold or reduce for creatinine > 1.5</td></tr>
<tr><td rowspan="4">FAM
fluorouracil
(5-FU)/
doxorubicin/
mitomycin C</td><td>5-FU</td><td>600 mg/M²</td><td>IV</td><td>days 1,8,29,36</td></tr>
<tr><td>Doxorubicin</td><td>30 mg/M²</td><td>IV</td><td>days 1,29</td></tr>
<tr><td>Mitomycin C</td><td>10 mg/M²</td><td>IV</td><td>day 1</td></tr>
<tr><td colspan="4">REF: Sporn et al. Semin Oncol 1993; 20:261-267

PREMEDICATIONS
1. Kytril 1 mg PO/IV 30 minutes before and 12 hours after therapy on days 1 and 29
2. Dexamethasone 20 mg IV 30 minutes before doxorubicin
3. Compazine 10 mg PO/IV 30 minutes before 5-FU PRN

Repeat every 56 days

Mitomycin C—myelosuppression occurs late (approximately 4 weeks); limit cumulative dose to 50 mg/M² (vascular toxicity)

Doxorubicin—monitor cumulative dose for cardiac toxicity (not to exceed 550 mg/M²); vesicant—avoid extravasation; use 50% for bilirubin 1.5-3.0 and 25% for bilirubin > 3.0</td></tr>
<tr><td rowspan="4">PCE
paclitaxel/
carboplatin/
etoposide
(VP-16)</td><td>Paclitaxel</td><td>200 mg/M²</td><td>IV (over 1 h)</td><td>day 1</td></tr>
<tr><td>Carboplatin</td><td>AUC 6</td><td>IV</td><td>day 1</td></tr>
<tr><td>VP-16</td><td>50 mg/
100 mg</td><td>PO alternating QOD</td><td>days 1-10</td></tr>
<tr><td colspan="4">REF: Hainsworth et al. J Clin Oncol 1997; 15:2385-2393</td></tr>
</table>

Continued

Agent	Dosage
	PREMEDICATIONS 1. Dexamethasone 20 mg IV 30 minutes before paclitaxel OR Dexamethasone 20 mg PO 6 and 12 hours prior to paclitaxel 2. Diphenhydramine 25-50 mg IV 30 minutes before paclitaxel 3. Cimetidine 300 mg IV 30 minutes before paclitaxel 4. Kytril 1 mg PO/IV 30 minutes before and 12 hours after carboplatin OTHER MEDICATIONS 1. Dexamethasone 4 mg PO BID for 6 doses after paclitaxel (for myalgias) 2. Give cisplatin delayed emesis prophylaxis Repeat every 21 days

Chapter 4
Endocrine Cancer

- Adrenocortical Carcinoma
- Carcinoid and Islet Cell Carcinoma
- Medullary Carcinoma of Thyroid
- Pheochromocytoma

Endocrine Cancer

Adrenocortical Carcinoma

Agent	Dosage			
CE cisplatin/ etoposide	Cisplatin	40 mg/M²	IV	days 1-3
	Etoposide	100 mg/M²	IV	days 1-3
	REF: Johnson et al. Cancer 1986; 58:2198-2202 PREMEDICATIONS 1. Kytril 1 mg PO/IV 30 minutes before and 12 hours after chemotherapy on days 1-3 2. Dexamethasone 20 mg IV 30 minutes before chemotherapy on days 1-3 OTHER MEDICATIONS 1. Give cisplatin delayed-emesis prophylaxis Repeat every 21 days Cisplatin—vigorous hydration is required; can be nephrotoxic and ototoxic; can cause peripheral neuropathy; hold or reduce for creatinine > 1.5			
CM cisplatin/ mitotane	Cisplatin	100 mg/M²	IV	day 1
	–dose reduced to 75 mg/M² in poor risk patients			
	Mitotane	1000 mg	PO	QID daily
	- advance dose as tolerated			
	REF: Bukowski et al. J Clin Oncol 1993; 11:161-165 PREMEDICATIONS 1. Kytril 1 mg PO/IV 30 minutes before and 12 hours after Cisplatin 2. Dexamethasone 20 mg IV 30 minutes before Cisplatin 3. Compazine 10 mg PO/IV 30 minutes before each dose of mitotane if needed OTHER MEDICATIONS 1. Give cisplatin delayed–emesis prophylaxis Repeat every 21days Cisplatin—vigorous hydration is required; can be nephrotoxic and ototoxic; can cause peripheral neuropathy; hold or reduce for creatinine > 1.5			

Continued

Agent	Dosage
	Mitotane—if well-tolerated, dose may be doubled on day 3; then, from day 5 onwards, may increase dose by 500 mg every 2-3 days until maximum tolerated dose (8-12 grams daily) has been reached; glucocorticoid and mineralocorticoid replacement necessary to prevent adrenal insufficiency; increased steroid doses may be needed at times of physiologic stress
MS mitotane/ streptozocin	Mitotane 2000-4000 mg PO QD (in 4 divided doses) Streptozocin 1000 mg IV days 1-5 –followed by 1500 to 2000 mg monthly maintenance REF: Eriksson et al. Cancer 1987; 59:1398-1403 PREMEDICATIONS 1. Kytril 1 mg PO/IV 30 minutes before and 12 hours after chemotherapy on days 1-5 2. Dexamethasone 10 mg IV 30 minutes before chemotherapy on days 1-5 OTHER MEDICATIONS 1. Give non-cisplatin delayed emesis prophylaxis Streptozocin—vesicant–avoid extravasation; have 50% dextrose available in case of sudden hypoglycemia; monitor closely for renal impairment Mitotane—if well-tolerated, dose may be doubled on day 3; then, from day 5 onwards, may increase dose by 500 mg every 2-3 days until maximum tolerated dose (8-12 grams daily) has been reached; glucocorticoid and mineralocorticoid replacement necessary to prevent adrenal insufficiency; increased steroid doses may be needed at times of physiologic stress
Mitotane (o.p.-DDD)	Mitotane 6-15 mg/kg PO QD (in 3-4 divided doses) REF: Wooten et al. Cancer 1993; 72:3145-3155 Mitotane—if well-tolerated, dose may be doubled on day 3; then, from day 5 onwards, may increase dose by 500 mg every 2-3 days until maximum tolerated dose (8-12 grams daily) has been reached; glucocorticoid and mineralocorticoid replacement necessary to prevent adrenal insufficiency; increased steroid doses may be needed at times of physiologic stress

Carcinoid and Islet Cell Carcinoma

Agent	Dosage			
CE cisplatin/ etoposide	Cisplatin	100 mg/M²	IV	day 1
	Etoposide	120 mg/M²	IV	day 1
	REF: Davis et al. Proc Am Soc Clin Oncol 1987; 6:73 PREMEDICATIONS 1. Kytril 1 mg PO/IV 30 minutes before and 12 hours after chemotherapy 2. Dexamethasone 20 mg IV 30 minutes before chemotherapy OTHER MEDICATIONS 1. Give cisplatin delayed-emesis prophylaxis Repeat every 21days Cisplatin—vigorous hydration is required; can be nephrotoxic and ototoxic; can cause peripheral neuropathy; hold or reduce for creatinine > 1.5			
Doxorubicin/ cisplatin	Doxorubicin	50 mg/M²	IV	day 1
	Cisplatin	50 mg/M²	IV	day 1
	REF: Sridhar et al. Cancer 1985; 55:2634-2637 PREMEDICATIONS 1. Kytril 1 mg PO/IV 30 minutes before and 12 hours after chemotherapy 2. Dexamethasone 20 mg IV 30 minutes before chemotherapy OTHER MEDICATIONS 1. Give cisplatin delayed-emesis prophylaxis Repeat every 21-28 days Cisplatin—vigorous hydration is required; can be nephrotoxic and ototoxic; can cause peripheral neuropathy; hold or reduce for creatinine > 1.5 Doxorubicin—monitor cumulative dose for cardiac toxicity (not to exceed 550 mg/M² or 450 mg/M² with prior chest radiotherapy); vesicant—avoid extravasation; use 50% for bilirubin 1.5-3.0; use 25% for bilirubin > 3.0			
Streptozocin/ doxorubicin	Streptozocin	500 mg/M²	IV	days 1-5
	Doxorubicin	50 mg/M²	IV	days 1, 22
	REF: Moertel et al. NEJM 1992; 326:519-523			

Continued

Agent	Dosage
	PREMEDICATIONS 1. Kytril 1 mg PO/IV 30 minutes before and 12 hours after chemotherapy on days 1-5 and 22 2. Dexamethasone 10 mg IV 30 minutes before chemotherapy on days 1-5 and 22 OTHER MEDICATIONS 1. Give non-cisplatin delayed emesis prophylaxis Repeat every 42 days Streptozocin—vesicant—avoid extravasation; have 50% dextrose available in case of sudden hypoglycemia; monitor closely for renal impairment Doxorubicin—monitor cumulative dose for cardiac toxicity (not to exceed 550 mg/M^2 or 450 mg/M^2 with prior chest radiotherapy); vesicant—avoid extravasation; use 50% for bilirubin 1.5-3.0; use 25% for bilirubin > 3.0
Streptozocin/ fluorouracil (5-FU)	Streptozocin 500 mg/M^2 IV days 1-5 5-FU 400 mg/M^2 IV days 1-5 REF: Moertel et al. NEJM 1980; 303:1189-1194 PREMEDICATIONS 1. Kytril 1 mg PO/IV 30 minutes before and 12 hours after chemotherapy on days 1-5 2. Dexamethasone 10 mg IV 30 minutes before chemotherapy on days 1-5 OTHER MEDICATIONS 1. Give non-cisplatin delayed emesis prophylaxis Repeat every 42 days Streptozocin—vesicant—avoid extravasation; have 50% dextrose available in case of sudden hypoglycemia; monitor closely for renal impairment

Medullary Carcinoma of Thyroid

Agent	Dosage
CVD cyclophosphamide/ vincristine/ dacarbazine (DTIC)	Cyclophosphamide 750 mg/M² IV day 1 Vincristine 1.4 mg/M² IV day 1 DTIC 600 mg/M² IV days 1, 2 REF: Wu et al. Cancer 1994; 73:432-436 PREMEDICATIONS 1. Kytril 1 mg PO/IV 30 minutes before and 12 hours after chemotherapy on days 1 and 2 2. Dexamethasone 20 mg IV 30 minutes before chemotherapy on days 1 and 2 Repeat every 21-28 days Dacarbazine—vesicant—avoid extravasation Vincristine—vesicant—avoid extravasation; cumulative neurotoxicity—may produce severe constipation; maximum 2 mg per administration
Dacarbazine (DTIC)/ fluorouracil (5-FU)	DTIC 250 mg/M² IV days 1-5 (over 15-30 min) 5-FU 450 mg/M² IV days 1-5 (over 12 hours) REF: Orlandi et al. Ann Oncol 1994; 5:763-765 PREMEDICATIONS 1. Kytril 1 mg PO/IV 30 minutes before and 12 hours after chemotherapy on days 1-5 2. Dexamethasone 10 mg IV 30 minutes before chemotherapy on days 1-5 Repeat every 28 days (maximum of 6 cycles) Dacarbazine—vesicant–avoid extravasation

Pheochromocytoma

Agent	Dosage
CVD cyclophosphamide/ vincristine/ dacarbazine (DTIC)	Cyclophosphamide 750 mg/M² IV day 1 Vincristine 1.4 mg/M² IV day 1 DTIC 600 mg/M² IV days 1, 2 REF: Averbuch et al. Ann Intern Med 1988; 109:267-273 PREMEDICATIONS 1. Kytril 1 mg PO/IV 30 minutes before and 12 hours after chemotherapy on days 1 and 2 2. Dexamethasone 20 mg IV 30 minutes before chemotherapy on days 1 and 2 Repeat every 21-28 days Dacarbazine—vesicant—avoid extravasation Vincristine—vesicant—avoid extravasation; cumulative neurotoxicity–may produce severe constipation; maximum 2 mg per administration

Chapter 5
Gastrointestinal Cancer

- Anal Cancer
- Colorectal Carcinoma
- Esophageal Cancer
- Gastric Carcinoma
- Pancreatic Cancer

Gastrointestinal Cancer

Anal Cancer

Agent	Dosage			
Fluorouracil (5-FU)/ mitomycin C/ XRT	5-FU	1000 mg/M²/d	CIV (X 4 days)	days 1-4 & 29-32
	Mitomycin C	10 mg/M²	IV	days 1,29

- maximum dose of mitomycin C is 20 mg

- given concurrently with XRT to 45 Gy over 5 weeks

If residual tumor is present on post-therapy biopsy:

5-FU	1000 mg/M²/d	CIV (X 4 days)	days 1-4
Cisplatin	100 mg/M²	IV	day 2

- given with XRT boost of 9 Gy over 5 days

REF: Flan et al. J Clin Oncol 1996; 14:2527-2539

PREMEDICATIONS
1. Kytril 1 mg PO/IV 30 minutes before and 12 hours after chemotherapy on days 1 and 29
2. Dexamethasone 20 mg IV 30 minutes before chemotherapy on days 1 and 29
3. If cisplatin is required, give above medications on day 2 before and after cisplatin

OTHER MEDICATIONS
1. Give cisplatin delayed-emesis prophylaxis (if cisplatin is required)

Mitomycin C—myelosuppression occurs late (approximately 4 weeks); limit cumulative dose to 50 mg/M² (vascular toxicity)

Cisplatin—vigorous hydration is required; can be nephrotoxic and ototoxic; can cause peripheral neuropathy; hold or reduce for creatinine > 1.5

Colorectal Carcinoma

Rectal Cancer

Agent	Dosage			
Fluorouracil (5-FU)/ radiotherapy	5-FU	500 mg/M²	IV bolus	days 1-5, 36-40
	5-FU	225 mg/M²/d	CIV	days 56-96
	- XRT 45 Gy given in 180 cGy fractions over 6 weeks starting day 56			
	5-FU	450 mg/M²	IV bolus	days 120-124, 134-138,169-173
	REF: O'Connell et al. NEJM 1994; 331:502-507			

Colorectal Cancer

Agent	Dosage			
Fluorouracil (5-FU)/ leucovorin (Mayo)– adjuvant	5-FU	425 mg/M²	IV bolus	days 1-5
	Leucovorin	20 mg/M²	IV bolus	days 1-5
	REF: O'Connell et al. J Clin Oncol 1997; 15:246-250			
	Repeat every 28 days for 6 cycles			
Fluorouracil (5-FU)/ leucovorin– adjuvant	Leucovorin	500 mg/M²	IV (over 2 h)	weekly for 6 wks
	followed 1 hour later by			
	5-FU	500 mg/M²	IV bolus	weekly for 6 wks
	REF: Wolmark et al. J Clin Oncol 1993; 11:1879-1887			
	Repeat every 56 days			
Fluorouracil (5-FU)/ levamisole- adjuvant	5-FU	450 mg/M²	IV	days 1-5
	then a 3 week rest followed by			
	5-FU	450 mg/M²	IV	weekly for 48 wks
	Levamisole	150 mg	PO	days 1-3 every 2 wks for 1 yr
	REF: Moertel et al. J Clin Oncol 1995; 13:2936-2943			
	Therapy lasts a total of 52 weeks			
Fluorouracil (5-FU)/ leucovorin (De Gramont)- metastatic	5-FU	1500-2000 mg/M²/d	CIV (for 48 h)	days 1-2
	Leucovorin	500 mg/M²	IV (over 2 h)	days 1-2
	REF: De Gramont et al. Eur J Cancer 1998; 34:619-626			
	Repeat every 14 days			

Agent	Dosage			
Fluorouracil (5-FU)/ leucovorin (Mayo)– metastatic	5-FU	425 mg/M²	IV bolus	days 1-5
	Leucovorin	20 mg/M²	IV bolus	days 1-5
	REF: Buroker et al. J Clin Oncol 1994; 12:14-20			
	Repeat every 28-35 days			
High-dose fluorouracil (5-FU)/ leucovorin– metastatic	5-FU	2600 mg/M²/day	CIV (X 24 h)	day 1
	Leucovorin	500 mg/M²	IV (over 1 h) before 5-FU	day 1
	REF: Weh et al. Ann Oncol 1994; 5:233-237			
	Repeat every 7 days for 6 weeks, then after a 2-week rest, repeat cycle			
Irinotecan/ fluorouracil (5-FU)/ leucovorin– metastatic	Irinotecan	125 mg/M²	IV (over 90 min)	day 1
	Leucovorin	20 mg/M²	IV	day 1
	5-FU	500 mg/M²	IV	day 1
	REF: Saltz et al. Proc Amer Soc Clin Oncol 1999; 18:abstract 898			
	PREMEDICATIONS 1. Kytril 1 mg PO/IV 30 minutes before and 12 hours after chemotherapy			
	OTHER MEDICATIONS 1. Lomotil 4 mg PO at first sign of any loose stool and 2 mg every 2 hours until formed stool			
	Repeat every 7 days for 4 weeks, followed by a 2 week break, then repeat			
Trimetrexate/ fluorouracil (5-FU)/ leucovorin– metastatic	Trimetrexate	110 mg/M²	IV	day 1
	Leucovorin	200 mg/M²	IV	day 2
	5-FU	500 mg/M²	IV	day 2
	- give 5-FU immediately after Leucovorin			
	Leucovorin	15 mg	PO Q6H for 7 doses	days 2,3
	- start 6 hours after 5-FU			
	REF: Blanke et al. J Clin Oncol 1997; 15:915-920			
	Repeat every 7 days for 6 weeks, followed by a 2 week break, then repeat			
Capecitabine	Capecitabine	2510 mg/M²/d	PO (divided BID)	days 1-14
	REF: Van Cutsem et al. J Clin Oncol 2000; 18:1337-1345			
	PREMEDICATIONS 1. Compazine 10 mg PO before chemotherapy prn			
	Repeat every 21 days			

Agent	Dosage
Fluorouracil continuous infusion–metastatic	5-FU 300 mg/M^2/d CIV daily REF: Lokich et al. J Clin Oncol 1989; 7:425-432 Treatment is continued until toxicity requires discontinuation or disease progression
Irinotecan (weekly)–metastatic	Irinotecan 125 mg/M^2 IV (over 90 min) days 1,8,15,22 REF: Pitot et al. J Clin Oncol 1997; 15:2910-2919 PREMEDICATIONS 1. Kytril 1 mg PO/IV 30 minutes before and 12 hours after chemotherapy OTHER MEDICATIONS 1. Lomotil 4 mg PO at first sign of any loose stool and 2 mg every 2 hours until formed stool Repeat every 42 days
Irinotecan–metastatic	Irinotecan 350 mg/M^2 IV (over 30 min) day 1 REF: Rougier et al. J Clin Oncol 1997; 15:251-260 PREMEDICATIONS 1. Kytril 1 mg PO/IV 30 minutes before and 12 hours after chemotherapy OTHER MEDICATIONS 1. Lomotil 4 mg PO at first sign of any loose stool and 2 mg every 2 hours until formed stool Repeat every 21 days
Oxaliplatin–metastatic	Oxaliplatin 130 mg/M^2 IV (over 2 h) day 1 REF: Becouarn et al. J Clin Oncol 1998; 16:2739-2744 PREMEDICATIONS 1. Kytril 1 mg PO/IV 30 minutes before and 12 hours after chemotherapy 2. Dexamethasone 20 mg IV 30 minutes before chemotherapy Repeat every 21 days Oxaliplatin—can cause peripheral neuropathy which is generally reversible with cessation of treatment

Esophageal Cancer

Concurrent Chemotherapy/Radiotherapy Regimens

Agent	Dosage
Fluorouracil (5-FU)/ cisplatin/XRT (Wayne State)	Cisplatin 75 mg/M² IV days 1,29,50,71 5-FU 1000 mg/M²/d CIV days 1-4,29-32, 50-53,71-74 - above is given concurrently with XRT 50 Gy over 5 weeks REF: Al-Sarraf et al. J Clin Oncol 1997; 15:277-284 PREMEDICATIONS 1. Kytril 1 mg PO/IV 30 minutes before and 12 hours after chemotherapy on days 1, 29, 50 and 71 2. Dexamethasone 20 mg IV 30 minutes before chemotherapy on days 1, 29, 50 and 71 OTHER MEDICATIONS 1. Give cisplatin delayed-emesis prophylaxis Cisplatin—vigorous hydration is required; can be nephrotoxic and ototoxic; can cause peripheral neuropathy; hold or reduce for creatinine > 1.5
Fluorouracil (5-FU)/ cisplatin/XRT (Johns Hopkins)	Cisplatin 26 mg/M²/d CIV days 1-5,26-30 5-FU 300 mg/M²/d CIV days 1-30 - above is given concurrently with XRT 44 Gy at 200 cGy daily - above is followed by esophagectomy when possible REF: Forastiere et al. Cancer J Sci Am 1997; 3:144-152 PREMEDICATIONS 1. Kytril 1 mg PO/IV 30 minutes before and 12 hours after chemotherapy on days 1-5, 26-30 2. Dexamethasone 10 mg IV 30 minutes before chemotherapy on days 1-5, 26-30 OTHER MEDICATIONS 1. Give cisplatin delayed-emesis prophylaxis Cisplatin—vigorous hydration is required; can be nephrotoxic and ototoxic; can cause peripheral neuropathy; hold or reduce for creatinine > 1.5
Fluorouracil (5-FU)/ cisplatin/XRT (North Carolina)	Cisplatin 100 mg/M² IV day 1 5-FU 1000 mg/M²/d CIV days 1-4,29-32 - above is given concurrently with XRT 45 Gy over 5 weeks - above is followed by esophagectomy when possible REF: Bates et al. J Clin Oncol 1996; 14:156-163

Continued

Agent	Dosage
	PREMEDICATIONS 1. Kytril 1 mg PO/IV 30 minutes before and 12 hours after chemotherapy on day 1 2. Dexamethasone 20 mg IV 30 minutes before chemotherapy on day 1 OTHER MEDICATIONS 1. Give cisplatin delayed-emesis prophylaxis Cisplatin—vigorous hydration is required; can be nephrotoxic and ototoxic; can cause peripheral neuropathy; hold or reduce for creatinine > 1.5
Chemotherapy Regimens	
CF cisplatin/ fluorouracil (5-FU)	Cisplatin 100 mg/M² IV day 1 5-FU 1000 mg/M²/d CIV X 5 days days 1-5 REF: Kies et al. Cancer 1987; 60:2156-2160 –there are multiple variations of this regimen PREMEDICATIONS 1. Kytril 1 mg 30 minutes before and 12 hours after chemotherapy on day 1 2. Dexamethasone 20 mg IV 30 minutes before chemotherapy on day 1 OTHER MEDICATIONS 1. Give cisplatin delayed-emesis prophylaxis Repeat every 28 days Cisplatin—vigorous hydration is required; can be nephrotoxic and ototoxic; can cause peripheral neuropathy; hold or reduce for creatinine > 1.5
CP carboplatin/ paclitaxel	Paclitaxel 200 mg/M² IV (over 3 h) day 1 –followed by Carboplatin AUC 5 IV day 1 REF: Philip et al. Semin Oncol 1997; 24(6 Supp 19):86-88 PREMEDICATIONS 1. Dexamethasone 20 mg IV 30 minutes before chemotherapy OR Dexamethasone 20 mg PO 6 and 12 hours prior 2. Diphenhydramine 50 mg IV 30 minutes before chemotherapy 3. Cimetidine 300 mg IV 30 minutes before chemotherapy 4. Kytril 1 mg PO/IV 30 minutes before and 12 hours after chemotherapy

Agent	Dosage
	OTHER MEDICATIONS 1. Dexamethasone 4 mg PO BID for 6 doses after paclitaxel (for myalgias) 2. Give cisplatin delayed emesis prophylaxis Repeat every 21 days
FAP fluorouracil (5-FU)/ doxorubicin/ cisplatin	5-FU 600 mg/M^2 IV days 1,8 Doxorubicin 30 mg/M^2 IV day 1 Cisplatin 75 mg/M^2 IV day 1 REF: Gisselbrecht et al. Cancer 1983; 52:974-977 PREMEDICATIONS 1. Kytril 1 mg PO/IV 30 minutes before and 12 hours after chemotherapy on day 1 2. Dexamethasone 20 mg IV 30 minutes before chemotherapy on day 1 OTHER MEDICATIONS 1. Give cisplatin delayed-emesis prophylaxis Repeat every 28 days Doxorubicin—monitor cumulative dose for cardiac toxicity (not to exceed 550 mg/M^2 or 450 mg/M^2 with prior chest Radiotherapy); vesicant – avoid extravasation; use 50% for bilirubin 1.5-3.0; use 25% for bilirubin > 3.0 Cisplatin—vigorous hydration is required; can be nephrotoxic and ototoxic; can cause peripheral neuropathy; hold or reduce for creatinine > 1.5
Irinotecan/ cisplatin	Irinotecan 65 mg/M^2 IV days 1, 8, 15, 22 Cisplatin 30 mg/M^2 IV days 1, 8, 15, 22 REF: Ilson et al. J Clin Oncol 1999; 17:3270-3275 PREMEDICATIONS 1. Kytril 1 mg PO/IV 30 minutes before and 12 hours after chemotherapy 2. Dexamethasone 20 mg IV 30 minutes before chemotherapy OTHER MEDICATIONS 1. Give cisplatin delayed-emesis prophylaxis Repeat every 28 days Cisplatin—vigorous hydration is required; can be nephrotoxic and ototoxic; can cause peripheral neuropathy; hold or reduce for creatinine > 1.5

Continued

Agent	Dosage
PCE paclitaxel/ cisplatin/ etoposide	Paclitaxel 50 mg/M² IV days 1,4,8,11,15, and 18 Cisplatin 15 mg/M² IV days 1,4,8,11,15, and 18 Etoposide 40 mg/M² IV days 1,4,8,11,15, and 18 REF: Lokich et al. Cancer 1999; 85:2347-2351 PREMEDICATIONS 1. Dexamethasone 20 mg IV 30 minutes before chemotherapy OR Dexamethasone 20 mg PO 6 and 12 hours prior 2. Diphenhydramine 50 mg IV 30 minutes before chemotherapy 3. Cimetidine 300 mg IV 30 minutes before chemotherapy 4. Kytril 1 mg PO/IV 30 minutes before and 12 hours after chemotherapy OTHER MEDICATIONS 1. Dexamethasone 4 mg PO BID for 6 doses after paclitaxel (for myalgias) 2. May need to give cisplatin delayed-emesis prophylaxis Repeat cycle every 28 days Cisplatin—vigorous hydration is required; can be nephrotoxic and ototoxic; can cause peripheral neuropathy; hold or reduce for creatinine > 1.5
TCF paclitaxel/ cisplatin/ fluorouracil (5-FU)	Paclitaxel 175 mg/M² IV (over 3 h) day 1 Cisplatin 20 mg/M² IV days 1-5 - dose is decreased to 15 mg/M² after 3rd cycle 5-FU 750 mg/M² IV days 1-5 REF: Ilson et al. J Clin Oncol 1998; 16:1826-1834 PREMEDICATIONS 1. Dexamethasone 20 mg IV 30 minutes before chemotherapy on days 1-5 2. Diphenhydramine 50 mg IV 30 minutes before chemotherapy on day 1 3. Cimetidine 300 mg IV 30 minutes before chemotherapy on day 1 4. Kytril 1 mg PO/IV 30 minutes before and 12 hours after chemotherapy on days 1-5 OTHER MEDICATIONS 1. Dexamethasone 4 mg PO BID for 6 doses after paclitaxel (for myalgias)

39

Agent	Dosage
	2. Give cisplatin delayed-emesis prophylaxis Repeat every 28 days Cisplatin—vigorous hydration is required; can be nephrotoxic and ototoxic; can cause peripheral neuropathy; hold or reduce for creatinine > 1.5
Paclitaxel	Paclitaxel 250 mg/M²CIV over 24 hours day 1 - studies are currently underway utilizing 80 mg/M² IV over 1 hour weekly REF: Ajani et al. Semin Oncol 1995; 22(3 Suppl 6):35-40 PREMEDICATIONS 1. Dexamethasone 20 mg IV 30 minutes before chemotherapy 2. Diphenhydramine 50 mg IV 30 minutes before chemotherapy 3. Cimetidine 300 mg IV 30 minutes before chemotherapy OTHER MEDICATIONS 1. Dexamethasone 4 mg PO BID for 6 doses after (for myalgias) 2. Requires use of G-CSF Repeat every 21 days

Brain Cancer
Breast Cancer
Carcinoma of Unknown Primary
Endocrine Cancer
Gastrointestinal Cancer
Genitourinary Cancer
Gynecologic Cancer
Head and Neck Cancer
Hematologic Malignancies

Gastric Carcinoma

Adjuvant Concurrent Chemo/Radiotherapy

Agent	Dosage
Fluorouracil (5-FU)/ leucovorin/ XRT–adjuvant	5-FU 425 mg/M² IV bolus days 1-5 Leucovorin 20 mg/M² IV bolus days 1-5 -above is given for 1 cycle postoperatively, followed by 5-FU 425 mg/M² IV bolus days 1-4,38-40 Leucovorin 20 mg/M² IV bolus days 1-4,38-40 -above is given concurrently with XRT to 4500 cGy in 180 cGy fractions -chemotherapy is given on first 4 and last 3 days of radiotherapy -this is followed by 5-FU 425 mg/M² IV bolus days 1-5 Leucovorin 20 mg/M² IV bolus days 1-5 -above portion of regimen is repeated every 28 days for 2 cycles post-concurrent therapy REF: MacDonald et al. Proc ASCO 2000: abstract 1

Chemotherapy for Advanced Disease

Agent	Dosage
EAP-2 etoposide (VP-16)/ doxorubicin/ cisplatin	VP-16 100 mg/M² IV days 1-3 Doxorubicin 40 mg/M² IV day 1 Cisplatin 25-30 mg/M² IV days 1-3 REF: Haim et al. Oncology 1994; 51:102-107 PREMEDICATIONS 1. Kytril 1 mg PO/IV 30 minutes before and 12 hours after chemotherapy on days 1-3 2. Dexamethasone 20 mg IV 30 minutes before chemotherapy on days 1-3 OTHER MEDICATIONS 1. Give cisplatin delayed-emesis prophylaxis Repeat every 21 days Cisplatin—vigorous hydration is required; can be nephrotoxic and ototoxic; can cause peripheral neuropathy; hold or reduce for creatinine > 1.5 Doxorubicin—monitor cumulative dose for cardiac toxicity (not to exceed 550 mg/M² or 450 mg/M² with prior chest radiotherapy); vesicant—avoid extravasation; use 50% for bilirubin 1.5-3.0; use 25% for bilirubin > 3.0

41

<table>
<tr><th>Agent</th><th>Dosage</th></tr>
<tr><td>ECF
epirubicin/
cisplatin/
fluorouracil
(5-FU)</td><td>Epirubicin 50 mg/M² IV day 1
Cisplatin 60 mg/M² IV day 1
5-FU 200 mg/M²/d CIV(X21 days) daily

REF: Webb et al. J Clin Oncol 1997; 15:261-267

PREMEDICATIONS
1. Kytril 1 mg PO/IV 30 minutes before and 12 hours after chemotherapy on day 1
2. Dexamethasone 10 mg IV 30 minutes before chemotherapy on day 1

OTHER MEDICATIONS
1. Give cisplatin delayed-emesis prophylaxis

Repeat every 21 days

Cisplatin—vigorous hydration is required; can be nephrotoxic and ototoxic; can cause peripheral neuropathy; hold or reduce for creatinine > 1.5

Epirubicin—monitor cumulative dose for cardiac toxicity (not to exceed 1000 mg/M²); vesicant—avoid extravasation</td></tr>
<tr><td>EFP
etoposide
(VP-16)/
fluorouracil
(5-FU)/cisplatin</td><td>VP-16 90 mg/M² IV (over 2 h) days 1,3,5
5-FU 900 mg/M²/d CIV (X 5 days) days 1-5
Cisplatin 20 mg/M² IV days 1-5

REF: Ajani et al. J Clin Oncol 1990; 8:1231-1238

PREMEDICATIONS
1. Kytril 1 mg PO/IV 30 minutes before and 12 hours after chemotherapy on days 1-5
2. Dexamethasone 10 mg IV 30 minutes before chemotherapy on days 1-5

OTHER MEDICATIONS
1. Give cisplatin delayed-emesis prophylaxis

Repeat every 28 days

Cisplatin—vigorous hydration is required; can be nephrotoxic and ototoxic; can cause peripheral neuropathy; hold or reduce for creatinine > 1.5</td></tr>
<tr><td>ELF
etoposide
(VP-16)/
leucovorin/
fluorouracil
(5-FU)</td><td>Leucovorin 300 mg/M² IV (over 10 min) days 1-3
VP-16 120 mg/M² IV (over 50 min) days 1-3
5-FU 500 mg/M² IV (over 10 min) days 1-3</td></tr>
</table>

Continued

Brain Cancer | Breast Cancer | Carcinoma of Unknown Primary | Endocrine Cancer | Gastrointestinal Cancer | Genitourinary Cancer | Gynecologic Cancer | Head and Neck Cancer | Hematologic Malignancies

Agent	Dosage
	REF: Wilke et al. Invest New Drugs 1990; 8:65-70 PREMEDICATIONS 1. Compazine 10 mg PO/IV 30 minutes before chemotherapy on days 1-3 Repeat every 21 days
FAM fluorouracil (5-FU)/ doxorubicin/ mitomycin C	5-FU 600 mg/M^2 IV days 1,8, 29,36 Doxorubicin 30 mg/M^2 IV days 1,29 Mitomycin C 10 mg/M^2 IV day 1 REF: MacDonald et al. Ann Intern Med 1980; 93:533-536 PREMEDICATIONS 1. Kytril 1 mg PO/IV 30 minutes before and 12 hours after chemotherapy on days 1 and 29 2. Dexamethasone 20 mg IV 30 minutes before chemotherapy on days 1 and 29 Repeat every 56 days Mitomycin C—myelosuppression occurs late (approximately 4 weeks); limit cumulative dose to 50 mg/M^2 (vascular toxicity) Doxorubicin—monitor cumulative dose for cardiac toxicity (not to exceed 550 mg/M^2 or 450 mg/M^2 with prior chest radiotherapy); vesicant—avoid extravasation; use 50% for bilirubin 1.5-3.0; use 25% for bilirubin > 3.0
FAMTx fluorouracil (5-FU)/ doxorubicin/ methotrexate	Methotrexate 1500 mg/M^2 IV day 1 - give MTX first and then wait 1 hour and give 5-FU 5-FU 1500 mg/M^2 IV day 1 Leucovorin 15 mg/M^2 PO Q6H days 2-4 - give total of 12 doses of Leucovorin, starting 24 hours after methotrexate Doxorubicin 30 mg/M^2 IV day 15 REF: Kelsen et al. J Clin Oncol 1992; 10:541-548 PREMEDICATIONS 1. Kytril 1 mg PO/IV 30 minutes before and 12 hours after chemotherapy on days 1 and 15 2. Dexamethasone 20 mg IV 30 minutes before chemotherapy on days 1 and 15 Repeat cycle on day 29 Doxorubicin—monitor cumulative dose for cardiac toxicity (not to exceed 550 mg/M^2 or 450 mg/M^2 with prior chest radiotherapy); vesicant—avoid extravasation; use 50% for bilirubin 1.5-3.0; use 25% for bilirubin > 3.0

Continued

Agent	Dosage
	Methotrexate—use 75% dose for CrCl < 50 and 50% dose if CrCl < 25; do not give if patient has an effusion ("reservoir effect")
Irinotecan/ cisplatin	Irinotecan 70 mg/M² IV days 1, 15 Cisplatin 80 mg/M² IV day 1 REF: Boku et al. J Clin Oncol 1999; 17:319-323 PREMEDICATIONS 1. Kytril 1 mg PO/IV 30 minutes before and 12 hours after chemotherapy on days 1 and 15 2. Dexamethasone 20 mg IV 30 minutes before chemotherapy on days 1 and 15 OTHER MEDICATIONS 1. Give cisplatin delayed-emesis prophylaxis 2. Lomotil 4 mg PO at first sign of any loose stool and 2 mg every 2 hours until formed stool Repeat every 28 days Cisplatin—vigorous hydration is required; can be nephrotoxic and ototoxic; can cause peripheral neuropathy; hold or reduce for creatinine > 1.5
PF paclitaxel/ fluorouracil (5-FU)	Paclitaxel 175 mg/M² IV (over 3 h) day 1 5-FU 1500 mg/M² IV (over 3 h) day 2 REF: Murad et al. Am J Clin Oncol 1999; 22:580-586 PREMEDICATIONS 1. Dexamethasone 20 mg IV 30 minutes before chemotherapy on day 1 2. Diphenhydramine 50 mg IV 30 minutes before chemotherapy on day 1 3. Cimetidine 300 mg IV 30 minutes before chemotherapy on day 1 4. Compazine 10 mg PO/IV 30 minutes before chemotherapy on day 2 OTHER MEDICATIONS 1. Dexamethasone 4 mg PO BID for 6 doses after paclitaxel (for myalgias) Repeat every 21 days for a maximum of 7 cycles
Fluorouracil (5-FU)	5-FU 500 mg/M² IV days 1-5 REF: Cullinan et al. J Clin Oncol 1994; 12:412-416 Repeat every 28 days

Pancreatic Cancer

Neoadjuvant Chemoradiation

Agent	Dosage
Fluorouracil (5-FU)/ mitomycin C/ XRT (ECOG)	5-FU 1000 mg/M²/d CIV days 2-5, 29-32 Mitomycin C 10 mg/M² IV day 2 - XRT given to 5040 cGy in 28 fractions starting on day 1 - definitive surgical resection follows completion of chemoradiation when possible REF: Hoffman et al. Am J Surg 1995; 169:71-77 PREMEDICATIONS 1. Kytril 1 mg PO/IV 30 minutes before and 12 hours after chemotherapy on day 2 2. Dexamethasone 20 mg IV 30 minutes before chemotherapy on day 2 Mitomycin C—myelosuppression occurs late (approximately 4 weeks); limit cumulative dose to 50 mg/M² (vascular toxicity)
Fluorouracil (5-FU)/XRT (MD Anderson)	5-FU 300 mg/M²/d CIV daily M-F during radiotherapy - XRT given to 5040 cGy in 28 fractions starting on day 1 - definitive surgical resection follows completion of chemoradiation when possible REF: Spitz et al. J Clin Oncol 1997; 15:928-937

Adjuvant Chemoradiation

Agent	Dosage
Fluorouracil (5-FU)/XRT	5-FU 500 mg/M² IV days 1-3,29-31 then weekly starting day 71 - weekly for 2 years (or until disease progression) - given concurrently with XRT, 20 Gy over 2 weeks, followed by a 2 week break, followed by an additional 20 Gy over 2 weeks - this regimen is utilized after maximal surgical resection REF: Gastrointestinal Tumor Study Group. Am Surg 1979; 189:205-208

Chemotherapy Regimens

Agent	Dosage			
SMF streptozocin/ mitomycin C/ fluorouracil (5-FU)	Streptozocin	1000 mg/M² over 1 hr	IV	days 1,8,29,36
	Mitomycin C	10 mg/M² bolus	IV	day 1
	5-FU	600 mg/M² bolus	IV	days 1,8,29,36
	REF: Wiggans et al. Cancer 1978; 41:387-391 PREMEDICATIONS 1. Kytril 1 mg PO/IV 30 minutes before and 12 hours after chemotherapy 2. Dexamethasone 20 mg IV 30 minutes before chemotherapy OTHER MEDICATIONS 1. Give non-cisplatin delayed emesis prophylaxis Repeat every 56 days Streptozocin—patients require aggressive hydration Mitomycin C—myelosuppression occurs late (approximately 4 weeks); limit cumulative dose to 50 mg/M² (vascular toxicity)			
Fluorouracil (5-FU)	5-FU	600 mg/M²	IV	weekly
	REF: Burris et al. J Clin Oncol 1997; 15:2403-2413			
Gemcitabine	Gemcitabine	1000 mg/M² over 30 min	IV	weekly X 7
	- followed by 1 week of rest; subsequent cycles given 3 weeks out of every 4 REF: Burris et al. J Clin Oncol 1997; 15:2403-2413 PREMEDICATIONS 1. Compazine 10 mg PO/IV 30 minutes before chemotherapy Repeat every 28 days (after 8 week induction course)			

Chapter 6
Genitourinary Cancer

- Bladder Cancer
- Penile Cancer
- Prostate Cancer
- Renal Cancer
- Testicular Cancer

Genitourinary Cancer

Bladder Cancer

Agent	Dosage			
Carboplatin/ paclitaxel	Paclitaxel	200 mg/M²	IV (over 3 h)	day 1
	–followed by			
	Carboplatin	AUC 5	IV (over 30 min)	day 1
	REF: Small et al. J Clin Oncol 2000; 18:2537-2544			
	PREMEDICATIONS 1. Dexamethasone 20 mg IV 30 minutes before chemotherapy OR Dexamethasone 20 mg PO 6 and 12 hours prior 2. Diphenhydramine 50 mg IV 30 minutes before chemotherapy 3. Cimetidine 300 mg IV 30 minutes before chemotherapy 4. Kytril 1 mg PO/IV 30 minutes before and 12 hours after chemotherapy			
	OTHER MEDICATIONS 1. Dexamethasone 4 mg PO BID for 6 doses after paclitaxel (for myalgias) 2. Give cisplatin delayed emesis prophylaxis			
	Repeat every 21 days			
CMV cisplatin/ methotrexate/ vinblastine	Cisplatin	100 mg/M²	IV	day 2
	Methotrexate	30 mg/M²	IV	days 1,8
	Vinblastine	4 mg/M²	IV	days 1,8
	REF: Jeffrey et al. Br J Cancer 1992; 66:542-546			
	PREMEDICATIONS 1. Kytril 1 mg PO/IV 30 minutes before and 12 hours after chemotherapy on day 2 2. Dexamethasone 20 mg IV 30 minutes before chemotherapy on day 2 3. Compazine 10 mg PO/IV 30 minutes before chemotherapy on days 1 and 8			
	OTHER MEDICATIONS 1. Give cisplatin delayed-emesis prophylaxis			
	Repeat every 21 days			

Continued

Agent	Dosage
	Cisplatin—vigorous hydration is required; can be nephrotoxic and ototoxic; can cause peripheral neuropathy; hold or reduce for creatinine > 1.5 Vinblastine—use 50% of dose for bilirubin > 3.0; vesicant–avoid extravasation; watch for neurotoxicity Methotrexate—use 75% dose for CrCl < 50; 50% dose if CrCl < 25; do not give if patient has an effusion ("reservoir effect")
Docetaxel/ cisplatin	Docetaxel 75 mg/M^2 IV (over 1 h) day 1 Cisplatin 75 mg/M^2 IV (over 1 h) day 1 REF: Dimopoulos et al. Ann Oncol 1999; 10:1385-1388 PREMEDICATIONS 1. Dexamethasone 20 mg IV 30 minutes before chemotherapy 2. Cimetidine 300 mg IV 30 minutes before chemotherapy 3. Diphenhydramine 25-50 mg IV 30 minutes before chemotherapy 4. Kytril 1 mg PO/IV 30 minutes before and 12 hours after chemotherapy OTHER MEDICATIONS 1. Dexamethasone 8 mg PO BID for 8 doses—start day prior to chemo (decreases lower extremity edema) 2. Give cisplatin delayed-emesis prophylaxis 3. G-CSF is given from day 5 until resolution of neutropenia Repeat every 21 days Cisplatin—vigorous hydration is required; can be nephrotoxic and ototoxic; can cause peripheral neuropathy; hold or reduce for creatinine > 1.5
Gemcitabine/ cisplatin	Gemcitabine 1000 mg/M^2 IV days 1,8,15 Cisplatin 70 mg/M^2 IV day 2 REF: Moore et al. J Clin Oncol 1999; 17:2876-2881 OR Gemcitabine 1000 mg/M^2 IV days 1,8,15 Cisplatin 75 mg/M^2 IV day 1 REF: Kaufman et al. J Clin Oncol 2000; 18:1921-1927 PREMEDICATIONS 1. Kytril 1 mg PO/IV 30 minutes before and 12 hours after chemotherapy on day 2 2. Dexamethasone 20 mg IV 30 minutes before chemotherapy on day 2 3. Compazine 10 mg PO/IV 30 minutes before chemotherapy on days 1, 8, and 15 if needed

Continued

Agent	Dosage
	OTHER MEDICATIONS 1. Give cisplatin delayed-emesis prophylaxis Repeat every 28 days Cisplatin—vigorous hydration is required; can be nephrotoxic and ototoxic; can cause peripheral neuropathy; hold or reduce for creatinine > 1.5
MVAC methotrexate/ vinblastine/ doxorubicin/ cisplatin	Methotrexate 30 mg/M² IV days 1,15,22 Vinblastine 3 mg/M² IV days 2,15,22 Doxorubicin 30 mg/M² IV day 2 - reduce dose to 15 mg/M² if patient has received > 2000 cGy in 5 days to pelvis Cisplatin 70 mg/M² IV day 2 - vinblastine and methotrexate given on days 15 and 22 only if WBC >2500 and platelet count is > 100,000 REF: Loehrer et al. J Clin Oncol 1992; 10:1066-1073 PREMEDICATIONS 1. Kytril 1 mg PO/IV 30 minutes before and 12 hours after chemotherapy on day 2 2. Dexamethasone 20 mg IV 30 minutes before chemotherapy on day 2 3. Compazine 10 mg PO/IV 30 minutes before chemotherapy on days 1, 15, and 22 if needed OTHER MEDICATIONS 1. Give cisplatin delayed-emesis prophylaxis Repeat every 28 days Cisplatin—vigorous hydration is required; can be nephrotoxic and ototoxic; can cause peripheral neuropathy; hold or reduce for creatinine > 1.5 Doxorubicin—monitor cumulative dose for cardiac toxicity (not to exceed 550 mg/M² or 450 mg/M² with prior chest radiotherapy); vesicant – avoid extravasation; use 50% for bilirubin 1.5-3.0; use 25% for bilirubin > 3.0 Vinblastine—use 50% of dose for bilirubin > 3.0; vesicant–avoid extravasation; watch for neurotoxicity Methotrexate – use 75% dose for CrCl < 50; 50% dose if CrCl < 25; do not give if patient has an effusion ("reservoir effect")

Agent	Dosage
Paclitaxel/ cisplatin	Paclitaxel 175 mg/M² IV (over 3 h) day 1 Cisplatin 75 mg/M² IV day 1 REF: Dreicer et al. J Clin Oncol 2000; 18:1058-1061 PREMEDICATIONS 1. Dexamethasone 20 mg IV 30 minutes before chemotherapy OR Dexamethasone 20 mg PO 6 and 12 hours prior 2. Diphenhydramine 50 mg IV 30 minutes before chemotherapy 3. Cimetidine 300 mg IV 30 minutes before chemotherapy 4. Kytril 1 mg PO/IV 30 minutes before and 12 hours after chemotherapy OTHER MEDICATIONS 1. Dexamethasone 4 mg PO BID for 6 doses after paclitaxel (for myalgias) 2. Give cisplatin delayed emesis prophylaxis Repeat every 21 days for a maximum of 6 cycles Cisplatin—vigorous hydration is required; can be nephrotoxic and ototoxic; can cause peripheral neuropathy; hold or reduce for creatinine > 1.5
Docetaxel	Docetaxel 100 mg/M² IV(over 1 h) day 1 REF: de Wit et al. Br J Cancer 1998; 78:1342-1345 PREMEDICATIONS 1. Dexamethasone 20 mg IV 30 minutes before chemotherapy 2. Cimetidine 300 mg IV 30 minutes before chemotherapy 3. Diphenhydramine 25-50 mg IV 30 minutes before chemotherapy OTHER MEDICATIONS 1. Dexamethasone 8 mg PO BID for 8 doses—start day prior to chemo (decreases lower extremity edema) Repeat every 21 days
Gemcitabine	Gemcitabine 1200 mg/M² IV days 1,8,15 REF: Moore et al. J Clin Oncol 1997; 15:3441-3445 PREMEDICATIONS 1. Compazine 10 mg PO/IV 30 minutes before chemotherapy on days 1, 8, and 15 Repeat every 28 days

Agent	Dosage
Paclitaxel	Paclitaxel 250 mg/M^2 CIV (over 24 h) day 1

REF: Roth et al. J Clin Oncol 1994; 12:2264-2270

PREMEDICATIONS
1. Dexamethasone 20 mg IV 30 minutes before chemo-therapy
OR
 Dexamethasone 20 mg PO 6 and 12 hours prior
2. Diphenhydramine 50 mg IV 30 minutes before chemo-therapy
3. Cimetidine 300 mg IV 30 minutes before chemotherapy

OTHER MEDICATIONS
1. Dexamethasone 4 mg PO BID for 6 doses after (for myalgias)
2. G-CSF support is required

Repeat every 21 days

Brain Cancer

Breast Cancer

Carcinoma of Unknown Primary

Endocrine Cancer

Gastrointestinal Cancer

Genitourinary Cancer

Gynecologic Cancer

Penile Cancer

Agent	Dosage
CF **cisplatin/** **fluorouracil** **(5-FU)**	Cisplatin 100 mg/M^2 IV day 1 5-FU 1000 mg/M^2/d (X 5 days) CIV days 1-5 REF: Shammas et al. J Urol 1992; 147:630-632 PREMEDICATIONS 1. Kytril 1 mg PO/IV 30 minutes before and 12 hours after chemotherapy on day 1 2. Dexamethasone 20 mg IV 30 minutes before chemotherapy on day 1 OTHER MEDICATIONS 1. Give cisplatin delayed-emesis prophylaxis Repeat every 21 days Cisplatin—vigorous hydration is required; can be nephrotoxic and ototoxic; can cause peripheral neuropathy; hold or reduce for creatinine > 1.5
MF **mitomycin C/** **fluorouracil** **(5-FU)**	Mitomycin C 10 mg/M^2 IV day 1 5-FU 1000 mg/M^2/d CIV days 1-4,29-32 - regimen is given concurrently with XRT REF: Oberfield et al. Br J Urol 1996; 78:573-578 PREMEDICATIONS 1. Kytril 1 mg PO/IV 30 minutes before and 12 hours after chemotherapy on day 1 2. Dexamethasone 20 mg IV 30 minutes before chemotherapy on day 1 Mitomycin C—myelosuppression occurs late (approximately 4 weeks); limit cumulative dose to 50 mg/M^2 (vascular toxicity)

Prostate Cancer

Hormonal Agents

LHRH agonists

Agent	Dosage
Goserelin (Zoladex)	3.6 mg SQ every 4 weeks REF: Soloway et al. Urology 1991; 37:46-51 10.8 mg SQ every 12 weeks REF: Dijkman et al. Eur Urol 1995; 27:43-46
Leuprolide (Lupron)	7.5 mg IM every 4 weeks REF: Leuprolide Study Group: NEJM 1984; 311:1281-1286 22.5 mg IM every 12 weeks REF: Sharifi et al. Clin Ther 1996; 18:647-657

Antiandrogens

Flutamide (Eulexin)	250 mg PO TID REF: McLeod et al. Prostate 1999; 40:218-224
Bicalutamide (Casodex)	50 mg PO QD REF: Schellhammer et al. Urology 1995; 45:745-752
Nilutamide (Nilandron)	150 mg PO QD REF: Janknegt et al. J Urol 1993; 149:77-82

Other Hormonal Agents

Aminoglute thimide/ hydrocortisone	Aminoglutethimide	250 mg	PO	QID
	Hydrocortisone	20 mg	PO	QAM
		10 mg	PO	QPM
	REF: Sartor et al. J Natl Cancer Inst 1994; 86:222-227			
Ketoconazole/ hydrocortisone	Ketoconazole	400 mg	PO	TID
	Hydrocortisone	20 mg	PO	QAM
		10 mg	PO	QPM
	REF: Small et al. Cancer 1997; 80:1755-1759			

Chemotherapy Regimens

Cyclophos-phamide/ etoposide (VP-16)	Cyclophosphamide	100 mg	PO	days 1-14
	VP-16	50 mg	PO	days 1-14
	REF: Maulard-Durdux et al. Cancer 1996; 77:1144-1148			
	Repeat every 28 days			

Agent	Dosage
Estramustine/ docetaxel	Estramustine 280 mg PO TID days 1-5 Docetaxel 70 mg/M² IV day 1 - dose is reduced to 60 mg/M² in extensively pretreated patients REF: Petrylak et al. J Clin Oncol 1999; 17:958-967 PREMEDICATIONS 1. Dexamethasone 20 mg IV 30 minutes before chemotherapy on day 1 2. Cimetidine 300 mg IV 30 minutes before chemotherapy on day 1 3. Diphenhydramine 25-50 mg IV 30 minutes before chemotherapy on day 1 4. Compazine 10 mg PO/IV 30 minutes before chemotherapy on day 1 OTHER MEDICATIONS 1. Dexamethasone 8 mg PO BID for 8 doses—start day prior to chemotherapy (decreases lower extremity edema) Repeat every 21 days
Estramustine/ etoposide (VP-16)	Estramustine 15 mg/kg/d PO (divided QID) days 1-21 VP-16 50 mg/M² PO (divided BID) days 1-21 REF: Pienta et al. J Clin Oncol 1994; 12:2005-2012 Repeat every 28 days
Estramustine/ etoposide (VP-16)/ paclitaxel	Estramustine 280 mg PO TID days 1-14 VP-16 100 mg PO days 1-14 Paclitaxel 135 mg/M² IV (over 1 h) day 2 REF: Smith et al. J Clin Oncol 1999; 17:1664-1674 PREMEDICATIONS 1. Dexamethasone 20 mg IV 30 minutes before chemotherapy on day 2 OR Dexamethasone 20 mg PO 6 and 12 hours prior 2. Diphenhydramine 50 mg IV 30 minutes before chemotherapy on day 2 3. Cimetidine 300 mg IV 30 minutes before chemotherapy on day 2 4. Kytril 1 mg PO/IV 30 minutes before and 12 hours after chemotherapy on day 2 OTHER MEDICATIONS 1. Dexamethasone 4 mg PO BID for 6 doses after paclitaxel (for myalgias) Repeat every 21 days

Agent	Dosage			
Estramustine/ vinblastine	Estramustine	600 mg/M^2	PO	days 1-42
	Vinblastine	4 mg/M^2	IV	days 1,8,15,22, 29,36
	REF: Hudes et al. J Clin Oncol 1999; 17:3160-3166			
	PREMEDICATIONS 1. Compazine 10 mg PO/IV 30 minutes before chemotherapy			
	Repeat every 56 days			
	Vinblastine—use 50% of dose for bilirubin > 3.0; vesicant–avoid extravasation; watch for neurotoxicity			
Mitoxantrone/ prednisone	Mitoxantrone	12 mg/M^2	IV	day 1
	Prednisone	5 mg	PO BID	daily
	REF: Tannock et al. J Clin Oncol 1996; 14:756-64			
	PREMEDICATIONS 1. Compazine 10 mg PO/IV prior to chemotherapy			
	Repeat every 21 days			
	Mitoxantrone—watch cumulative dose–do not exceed 140 mg/M^2; possible cardiac toxicity			
Cyclophos-phamide	Cyclophosphamide	100 mg/M^2	PO	days 1-14
	REF: Raghavan et al. Br J Urol 1993; 72:625-628			
	Repeat every 28 days			

Renal Cell Cancer

Agent	Dosage
Vinblastine/ interferon-alpha2a (IFN)	Vinblastine 0.1 mg/kg IV day 1 IFN 3 MIU SQ TIW week one –then IFN 18 MIU SQ TIW REF: Pyrhönen et al. J Clin Oncol 1999; 17:2859-2867 PREMEDICATIONS 1. Compazine 10 mg PO/IV 30 minutes before chemotherapy 2. Tylenol 650 mg PO prior to IFN Vinblastine—use 50% of dose for bilirubin > 3.0; vesicant–avoid extravasation; watch for neurotoxicity
Interferon (IFN)/ interleukin-2 (IL-2)	IL-2 4 MIU SQ days 1-4, weekly X 4 IFN 9 MIU SQ days 1-4, weekly X 4 REF: Vogelzang et al. J Clin Oncol 1993; 11:1809-1816 PREMEDICATIONS 1. Compazine 10 mg PO/IV 30 minutes before biotherapy 2. Tylenol 650 mg PO before biotherapy Repeat every 42 days
Alpha-interferon (IFN)	IFN 18 MIU IM TIW REF: Fossa et al. Ann Oncol 1992; 3:301-305 PREMEDICATIONS 1. Tylenol 650 mg PO prior to IFN
Interleukin-2 (IL-2) high-dose	IL-2 600,000-720,000 IU/kg IV (over 15 min) Q8h X 14 doses Repeat above in 6-9 days REF: Fyfe et al. J Clin Oncol 1995; 13:688-696 PREMEDICATIONS 1. Kytril 1 mg PO/IV 30 minutes before therapy and Q12H during therapy 2. Tylenol 650 mg PO 30 minutes before each dose of IL-2, and Q4H prn 3. Cimetidine 800 mg PO/IV daily during IL-2 therapy (given in single or divided doses) Repeat every 6-12 weeks IL-2 may cause capillary leak syndrome with profound hypotension and patients may require vasopressor support and aggressive fluid management. Patients should be cared for in an intensive care setting

Agent	Dosage
Interleukin-2 (IL-2)–low-dose	IL-2 3 MIU SQ BID days 1-5 weekly for 6 wks REF: Stadler et al. Semin Oncol 1995; 22:67-73 PREMEDICATIONS 1. Tylenol 650 mg PO 30 minutes before IL-2 daily 2. Compazine 10 mg PO 30 minutes before IL-2
Vinblastine	Vinblastine 0.1 mg/kg IV day 1 REF: Pyrhönen et al. J Clin Oncol 1999; 17:2859-2867 PREMEDICATIONS 1. Compazine 10 mg PO/IV 30 minutes before chemotherapy Repeat every 7 days Vinblastine—use 50% of dose for bilirubin > 3.0; vesicant–avoid extravasation; watch for neurotoxicity

Testicular Cancer

Agent	Dosage
BEP bleomycin/ etoposide (VP-16)/ cisplatin	Cisplatin 20 mg/M² IV days 1-5 VP-16 100 mg/M² IV days 1-5 Bleomycin 30 units IV days 2,9,16 REF: Einhorn et al. J Clin Oncol 1989; 7:387-391 PREMEDICATIONS 1. Kytril 1 mg PO/IV 30 minutes before and 12 hours after chemotherapy on days 1-5 2. Dexamethasone 10 mg IV 30 minutes before chemotherapy on days 1-5 OTHER MEDICATIONS 1. Give cisplatin delayed-emesis prophylaxis Repeat every 21 days Cisplatin—vigorous hydration is required; can be nephrotoxic and ototoxic; can cause peripheral neuropathy; hold or reduce for creatinine > 1.5 Bleomycin—give test dose of 1-2 units because of possible acute pulmonary, anaphylactoid, or severe febrile reactions; must dose adjust for renal insufficiency; total lifetime dose should not exceed 400 units; avoid high FiO_2 as it can exacerbate pulmonary toxicity
EP etoposide (VP-16)/ cisplatin	VP-16 100 mg/M² IV days 1-5 Cisplatin 20 mg/M² IV days 1-5 REF: Motzer et al. J Clin Oncol 1995; 13:2700-2704 PREMEDICATIONS 1. Kytril 1 mg PO/IV 30 minutes before and 12 hours after chemotherapy on days 1-5 2. Dexamethasone 10 mg IV 30 minutes before chemotherapy on days 1-5 OTHER MEDICATIONS 1. Give cisplatin delayed-emesis prophylaxis Repeat every 21 days Cisplatin—vigorous hydration is required; can be nephrotoxic and ototoxic; can cause peripheral neuropathy; hold or reduce for creatinine > 1.5

Agent	Dosage			
PVB cisplatin/ vinblastine/ bleomycin	Cisplatin	20.00 mg/M²	IV	days 1-5
	Vinblastine	0.15 mg/kg	IV	days 1, 2
	–reduce dose by 20% for prior radiotherapy			
	Bleomycin	30 units	IV	days 2,9,16
	REF: Einhorn et al. Ann Intern Med 1977; 87:293-298 PREMEDICATIONS 1. Kytril 1 mg PO/IV 30 minutes before and 12 hours after chemotherapy on days 1-5 2. Dexamethasone 10 mg IV 30 minutes before chemotherapy on days 1-5 OTHER MEDICATIONS 1. Give cisplatin delayed-emesis prophylaxis Repeat every 21 days Cisplatin—vigorous hydration is required; can be nephrotoxic and ototoxic; can cause peripheral neuropathy; hold or reduce for creatinine > 1.5 Bleomycin—give test dose of 1-2 units because of possible acute pulmonary, anaphylactoid, or severe febrile reactions; must dose adjust for renal insufficiency; total lifetime dose should not exceed 400 units; avoid high FiO_2 as it can exacerbate pulmonary toxicity Vinblastine—use 50% of dose for bilirubin > 3.0; vesicant–avoid extravasation; watch for neurotoxicity			
VeIP vinblastine/ ifosfamide/ cisplatin (salvage)	Vinblastine	0.11 mg/kg	IV	days 1,2
	Ifosfamide	1200 mg/M²/d	CIV (120 hr)	days 1-5
	Cisplatin	20 mg/M²	IV	days 1-5
	Mesna	400 mg/M²	IV	day 1
	–give bolus 15 minutes prior to Ifosfamide			
	Mesna	1200 mg/M²/d	CIV (120 hr)	days 1-5
	–start immediately after Mesna bolus REF: Loehrer et al. Ann Intern Med 1988; 109:540-546 PREMEDICATIONS 1. Kytril 1 mg PO/IV 30 minutes before and 12 hours after chemotherapy on days 1-5 2. Dexamethasone 10 mg IV 30 minutes before chemotherapy on days 1-5 OTHER MEDICATIONS 1. Give cisplatin delayed-emesis prophylaxis Repeat every 21 days			

Continued

Agent	Dosage
	Cisplatin—vigorous hydration is required; can be nephrotoxic and ototoxic; can cause peripheral neuropathy; hold or reduce for creatinine > 1.5 Ifosfamide—adequate hydration is necessary to prevent nephrotoxicity Vinblastine—use 50% of dose for bilirubin > 3.0; vesicant–avoid extravasation; watch for neurotoxicity
VIP etoposide (VP-16)/ ifosfamide/ cisplatin (salvage)	VP-16 75 mg/M^2 IV days 1-5 Ifosfamide 1200 mg/M^2/d CIV (120 hr) days 1-5 Cisplatin 20 mg/M^2 IV days 1-5 Mesna 400 mg/M^2 IV day 1 –give bolus 15 minutes prior to Ifosfamide Mesna 1200 mg/M^2/d CIV (120 hr) days 1-5 –start immediately after Mesna bolus REF: Loehrer et al. Ann Intern Med 1988; 109:540-546 PREMEDICATIONS 1. Kytril 1 mg PO/IV 30 minutes before and 12 hours after chemotherapy on days 1-5 2. Dexamethasone 10 mg IV 30 minutes before chemotherapy on days 1-5 OTHER MEDICATIONS 1. Give cisplatin delayed-emesis prophylaxis Repeat every 21 days Cisplatin—vigorous hydration is required; can be nephrotoxic and ototoxic; can cause peripheral neuropathy; hold or reduce for creatinine > 1.5 Ifosfamide—adequate hydration is necessary to prevent nephrotoxicity
Gemcitabine	–for use in refractory, heavily pretreated patients Gemcitabine 1200 mg/M^2 IV days 1,8,15 REF: Einhorn et al. J Clin Oncol 1999; 17:509-511 PREMEDICATIONS 1. Compazine 10 mg PO/IV 30 minutes before chemotherapy on days 1, 8, and 15 Repeat every 28 days

Chapter 7
Gynecologic Cancer

- Cervical Cancer
- Endometrial Cancer
- Ovarian Cancer
- Trophoblastic Cancer

Gynecologic Cancer

Cervical Cancer

Agent	Dosage			
BIP #1 bleomycin/ ifosfamide/ cisplatin	Bleomycin	15 mg	IV	day 1
	Ifosfamide	1000 mg/M²	IV	days 1-5
	Cisplatin	50 mg/M²	IV	day 1
	Mesna	1000 mg/M²	IV	days 1-5
	REF: Kumar et al. Gynecol Oncol 1991; 40:107-111 PREMEDICATIONS 1. Kytril 1 mg PO/IV 30 minutes before and 12 hours after chemotherapy on days 1-5 2. Dexamethasone 10 mg IV 30 minutes before chemotherapy on days 1-5 OTHER MEDICATIONS 1. Give cisplatin delayed-emesis prophylaxis Repeat every 21 days Cisplatin—vigorous hydration is required; can be nephrotoxic and ototoxic; can cause peripheral neuropathy; hold or reduce for creatinine > 1.5 Bleomycin—give test dose of 1-2 units because of possible acute pulmonary, anaphylactoid, or severe febrile reactions; must dose adjust for renal insufficiency; total lifetime dose should not exceed 400 units; avoid high FiO_2 as it can exacerbate pulmonary toxicity Ifosfamide—adequate hydration is necessary to prevent nephrotoxicity			
BIP #2 bleomycin/ ifosfamide/ cisplatin	Bleomycin	30 units	CIV (over 24 h)	day 1
	Ifosfamide	5000 mg/M²	CIV (over 24 h)	day 2
	Mesna	8000 mg/M²	CIV (over 36 h)	day 2
	–starting with ifosfamide			
	Cisplatin	50 mg/M²	IV	day 2
	REF: Buxton et al. J Natl Cancer Inst 1989; 81:359-361 PREMEDICATIONS 1. Kytril 1 mg PO/IV 30 minutes before and 12 hours after chemotherapy on days 1 and 2 2. Dexamethasone 10 mg IV 30 minutes before chemotherapy on days 1 and 2			

Continued

Agent	Dosage
	OTHER MEDICATIONS 1. Give cisplatin delayed-emesis prophylaxis Repeat every 21 days Cisplatin—vigorous hydration is required; can be nephrotoxic and ototoxic; can cause peripheral neuropathy; hold or reduce for creatinine > 1.5 Bleomycin—give test dose of 1-2 units because of possible acute pulmonary, anaphylactoid, or severe febrile reactions; must dose adjust for renal insufficiency; total lifetime dose should not exceed 400 units; avoid high FiO_2 as it can exacerbate pulmonary toxicity Ifosfamide—adequate hydration is necessary to prevent nephrotoxicity
Cisplatin/XRT (neoadjuvant)	Cisplatin 40 mg/M² IV weekly X 6 –given concurrently with XRT REF: Keys et al. NEJM 1999; 340:1154-1161 PREMEDICATIONS 1. Kytril 1 mg PO/IV 30 minutes before and 12 hours after chemotherapy 2. Dexamethasone 10 mg IV 30 minutes before chemotherapy OTHER MEDICATIONS 1. Give cisplatin delayed-emesis prophylaxis Cisplatin—vigorous hydration is required; can be nephrotoxic and ototoxic; can cause peripheral neuropathy; hold or reduce for creatinine > 1.5
Gemcitabine/ cisplatin	Gemcitabine 1250 mg/M² IV days 1, 8 Cisplatin 50 mg/M² IV day 1 REF: Burnett et al. Gynecol Oncol 2000; 76:63-66 PREMEDICATIONS 1. Kytril 1 mg PO/IV 30 minutes before and 12 hours after chemotherapy on day 1 2. Dexamethasone 10 mg IV 30 minutes before chemotherapy on day 1 OTHER MEDICATIONS 1. Give cisplatin delayed-emesis prophylaxis Repeat every 21 days Cisplatin—vigorous hydration is required; can be nephrotoxic and ototoxic; can cause peripheral neuropathy; hold or reduce for creatinine > 1.5

Agent	Dosage			
Paclitaxel/ cisplatin	Paclitaxel Cisplatin	175 mg/M² 75 mg/M²	IV (over 3 h) IV	day 1 day 1
	REF: Papadimitriou et al. J Clin Oncol 1999; 17:761-766 PREMEDICATIONS 1. Dexamethasone 20 mg IV 30 minutes before chemotherapy OR Dexamethasone 20 mg PO 6 and 12 hours prior to chemotherapy 2. Diphenhydramine 50 mg IV 30 minutes before chemotherapy 3. Cimetidine 300 mg IV 30 minutes before chemotherapy 4. Kytril 1 mg PO/IV 30 minutes before and 12 hours after chemotherapy OTHER MEDICATIONS 1. Dexamethasone 4 mg PO BID for 6 doses after paclitaxel (for myalgias) 2. Give cisplatin delayed-emesis prophylaxis Repeat every 28 days Cisplatin—vigorous hydration is required; can be nephrotoxic and ototoxic; can cause peripheral neuropathy; hold or reduce for creatinine > 1.5			
Docetaxel	Docetaxel	100 mg/M²	IV (over 1 h)	day 1
	REF: Kudelka et al. Anticancer Drugs 1996; 7:398-401 PREMEDICATIONS 1. Dexamethasone 20 mg IV 30 minutes before chemotherapy 2. Cimetidine 300 mg IV 30 minutes before chemotherapy 3. Diphenhydramine 25-50 mg IV 30 minutes before chemotherapy OTHER MEDICATIONS 1. Dexamethasone 8 mg PO BID for 8 doses—start day prior to chemotherapy (decreases lower extremity edema) Repeat every 21 days			
Irinotecan	Irinotecan	125 mg/M²	IV (over 90 min)	days 1,8, 15,22
	REF: Look et al. Gynecol Oncol 1998; 70:334-338 PREMEDICATIONS 1. Kytril 1 mg PO/IV 30 minutes before and 12 hours after chemotherapy			

Continued

Agent	Dosage
	OTHER MEDICATIONS 1. Lomotil 4 mg PO at first sign of any loose stool and 2 mg every 2 hours until formed stool Repeat every 42 days
Paclitaxel	Paclitaxel 250 mg/M² IV (over 3 hr) day 1 REF: Kudelka et al. Anticancer Drugs 1997; 8:657-661 PREMEDICATIONS 1. Dexamethasone 20 mg IV 30 minutes before chemotherapy OR Dexamethasone 20 mg PO 6 and 12 hours prior to chemotherapy 2. Diphenhydramine 50 mg IV 30 minutes before chemotherapy 3. Cimetidine 300 mg IV 30 minutes before chemotherapy OTHER MEDICATIONS 1. Dexamethasone 4 mg PO BID for 6 doses after (for myalgias) 2. Requires G-CSF support Repeat every 21 days

Endometrial Carcinoma

Agent	Dosage			
CAP cyclophos-phamide/ doxorubicin/ cisplatin	Cyclophosphamide	500 mg/M²	IV	day 1
	Doxorubicin	50 mg/M²	IV	day 1
	Cisplatin	50 mg/M²	IV	day 1

REF: Burke et al. Gynecol Oncol 1991; 40:264-267

PREMEDICATIONS
1. Kytril 1 mg PO/IV 30 minutes before and 12 hours after chemotherapy
2. Dexamethasone 20 mg IV 30 minutes before chemo-therapy

OTHER MEDICATIONS
1. Give cisplatin delayed-emesis prophylaxis

Repeat every 28 days

Cisplatin—vigorous hydration is required; can be nephrotoxic and ototoxic; can cause peripheral neuropathy; hold or reduce for creatinine > 1.5

Doxorubicin—monitor cumulative dose for cardiac toxicity (not to exceed 550 mg/M² or 450 mg/M² with prior chest radiotherapy); vesicant—avoid extravasation; use 50% for bilirubin 1.5-3.0; use 25% for bilirubin > 3.0

Agent	Dosage			
CP carboplatin/ paclitaxel	Paclitaxel	175 mg/M²	IV (over 3 h)	day 1
	followed by			
	Carboplatin	AUC 5	IV (over 1 h)	day 1

REF: Price et al. Semin Oncol 1997; 24(5suppl15):S78-S82

PREMEDICATIONS
1. Dexamethasone 20 mg IV 30 minutes before chemo-therapy
   OR
   Dexamethasone 20 mg PO 6 and 12 hours prior to chemotherapy
2. Diphenhydramine 50 mg IV 30 minutes before chemo-therapy
3. Cimetidine 300 mg IV 30 minutes before chemotherapy
4. Kytril 1 mg PO/IV 30 minutes before and 12 hours after chemotherapy

OTHER MEDICATIONS
1. Dexamethasone 4 mg PO BID for 6 doses after paclitaxel (for myalgias)
2. Give cisplatin delayed emesis prophylaxis

Repeat every 28 days

Brain Cancer | Breast Cancer | Carcinoma of Unknown Primary | Endocrine Cancer | Gastrointestinal Cancer | Genitourinary Cancer | Gynecologic Cancer | Head and Neck Cancer | Hematologic Malignancies

Agent	Dosage
Doxorubicin/ cisplatin	Doxorubicin 50 mg/M² IV day 1 Cisplatin 50 mg/M² IV day 1 REF: Deppe et al. Eur J Gynaecol Oncol 1994; 15:263-266 PREMEDICATIONS 1. Kytril 1 mg PO/IV 30 minutes before and 12 hours after chemotherapy 2. Dexamethasone 20 mg IV 30 minutes before chemotherapy OTHER MEDICATIONS 1. Give cisplatin delayed-emesis prophylaxis Repeat every 21 days Cisplatin—vigorous hydration is required; can be nephrotoxic and ototoxic; can cause peripheral neuropathy; hold or reduce for creatinine > 1.5 Doxorubicin—monitor cumulative dose for cardiac toxicity (not to exceed 550 mg/M² or 450 mg/M² with prior chest radiotherapy); vesicant—avoid extravasation; use 50% for bilirubin 1.5-3.0; use 25% for bilirubin > 3.0
Doxorubicin/ cyclophos-phamide	Doxorubicin 60 mg/M² IV day 1 Cyclophosphamide 500 mg/M² IV day 1 REF: Thigpen et al. J Clin Oncol 1994; 12:1408-1414 PREMEDICATIONS 1. Kytril 1 mg PO/IV 30 minutes before and 12 hours after chemotherapy 2. Dexamethasone 20 mg IV 30 minutes before chemotherapy Repeat every 21 days Doxorubicin—monitor cumulative dose for cardiac toxicity (not to exceed 550 mg/M² or 450 mg/M² with prior chest radiotherapy); vesicant—avoid extravasation; use 50% for bilirubin 1.5-3.0; use 25% for bilirubin > 3.0
Medroxy-progesterone	Medroxyprogesterone 200 mg PO daily REF: Thigpen et al. J Clin Oncol 1999; 17:1736-1744
Paclitaxel	Paclitaxel 175 mg/M² IV over 3 hours day 1 REF: Lissoni et al. Ann Oncol 1996; 7:861-863

Continued

Agent	Dosage
	PREMEDICATIONS 1. Dexamethasone 20 mg IV 30 minutes before chemotherapy OR Dexamethasone 20 mg PO 6 and 12 hours prior to chemotherapy 2. Diphenhydramine 50 mg IV 30 minutes before chemotherapy 3. Cimetidine 300 mg IV 30 minutes before chemotherapy OTHER MEDICATIONS 1. Dexamethasone 4 mg PO BID for 6 doses after (for myalgias) Repeat every 21 days

Ovarian Cancer

Agent	Dosage
CC **carboplatin/ cyclophos-phamide**	Carboplatin 300 mg/M^2 IV day 1 Cyclophosphamide 600 mg/M^2 IV day 1 REF: Alberts et al. J Clin Oncol 1992; 10:706-717 PREMEDICATIONS 1. Kytril 1 mg IV/PO 30 minutes before and 12 hours after chemotherapy 2. Dexamethasone 20 mg IV 30 minutes before chemo-therapy OTHER MEDICATIONS 1. Give cisplatin delayed emesis prophylaxis Repeat every 28 days
CP **cyclophos-phamide/ cisplatin**	Cyclophosphamide 600 mg/M^2 IV day 1 Cisplatin 100 mg/M^2 IV day 1 REF: Alberts et al. J Clin Oncol 1992; 10:706-717 OR Cyclophosphamide 750 mg/M^2 IV day 1 Cisplatin 75 mg/M^2 IV day 1 REF: McGuire et al. NEJM 1996; 334:1-6 PREMEDICATIONS 1. Kytril 1 mg PO/IV 30 minutes before and 12 hours after chemotherapy 2. Dexamethasone 20 mg IV 30 minutes before chemo-therapy OTHER MEDICATIONS 1. Give cisplatin delayed-emesis prophylaxis Repeat every 21 days Cisplatin—vigorous hydration is required; can be nephrotoxic and ototoxic; can cause peripheral neuropathy; hold or reduce for creatinine > 1.5
CT **paclitaxel/ cisplatin**	Paclitaxel 135 mg/M^2 CIV (over 24 h) day 1 Cisplatin 75 mg/M^2 IV day 1 REF: McGuire et al. NEJM 1996; 334:1-6

Continued

Agent	Dosage
	OR Paclitaxel 175 mg/M² IV (over 3 h) day 1 Cisplatin 75 mg/M² IV day 1 REF: Piccant et al. Proc ASCO 1997; 16:abstract 1258 PREMEDICATIONS 1. Dexamethasone 20 mg IV 30 minutes before chemotherapy OR Dexamethasone 20 mg PO 6 and 12 hours prior to chemotherapy 2. Diphenhydramine 50 mg IV 30 minutes before chemotherapy 3. Cimetidine 300 mg IV 30 minutes before chemotherapy 4. Kytril 1 mg PO/IV 30 minutes before and 12 hours after chemotherapy OTHER MEDICATIONS 1. Give cisplatin delayed-emesis prophylaxis 2. Dexamethasone 4 mg PO BID for 6 doses after paclitaxel (for myalgias) Repeat every 21 days Cisplatin—vigorous hydration is required; can be nephrotoxic and ototoxic; can cause peripheral neuropathy; hold or reduce for creatinine > 1.5
PC paclitaxel/ carboplatin	Paclitaxel 175 mg/M² IV (over 3 h) day 1 followed by Carboplatin AUC 7-7.5 IV (over 1 h) day 1 REF: Coleman et al. Cancer J Sci Am 1997; 3:246-253 PREMEDICATIONS 1. Dexamethasone 20 mg IV 30 minutes before chemotherapy OR Dexamethasone 20 mg PO 6 and 12 hours prior to chemotherapy 2. Diphenhydramine 50 mg IV 30 minutes before chemotherapy 3. Cimetidine 300 mg IV 30 minutes before chemotherapy 4. Kytril 1 mg PO/IV 30 minutes before and 12 hours after chemotherapy OTHER MEDICATIONS 1. Dexamethasone 4 mg PO BID for 6 doses after paclitaxel (for myalgias) 2. Give cisplatin delayed emesis prophylaxis Repeat every 21 days

Agent	Dosage
Docetaxel	Docetaxel 100 mg/M² IV (over 1 h) day 1 REF: Kaye et al. Eur J Cancer 1997; 33:2167-2170 PREMEDICATIONS 1. Dexamethasone 20 mg IV 30 minutes before chemotherapy 2. Cimetidine 300 mg IV 30 minutes before chemotherapy 3. Diphenhydramine 25-50 mg IV 30 minutes before chemotherapy 4. Compazine 10 mg PO/IV 30 minutes before chemotherapy OTHER MEDICATIONS 1. Dexamethasone 8 mg PO BID for 8 doses—start day prior to chemo (decreases lower extremity edema) Repeat every 21 days
Etoposide (VP-16)	VP-16 50 mg PO BID days 1-7 REF: de Jong et al. Gynecol Oncol 1997; 66:197-201 Repeat every 21 days
Gemcitabine	Gemcitabine 1250 mg/M² IV days 1,8,15 REF: von Minckwitz et al. Ann Oncol 1999; 10:853-855 PREMEDICATIONS 1. Compazine 10 mg PO/IV 30 minutes before Repeat every 28 days
Hexamethyl-melamine	Hexamethylmelamine 260 mg/M²/d PO days 1-14 REF: Markman et al. Gynecol Oncol 1998: 69:226-229 Repeat every 28 days Hexamethylmelamine—can have dose-limiting nausea and vomiting
Liposomal doxorubicin (Doxil)	Doxil 50 mg/M² IV day 1 REF: Muggia et al. J Clin Oncol 1997; 15:987-993 PREMEDICATIONS 1. Kytril 1 mg PO/IV 30 minutes before and 12 hours after chemotherapy 2. Dexamethasone 20 mg IV 30 minutes before chemotherapy Repeat every 21-28 days

Continued

Agent	Dosage
	Doxorubicin—monitor cumulative dose for cardiac toxicity (not to exceed 550 mg/M² or 450 mg/M² with prior chest radiotherapy); vesicant—avoid extravasation; use 50% for bilirubin 1.5-3.0; use 25% for bilirubin > 3.0
Paclitaxel	Paclitaxel 175 mg/M² IV (over 3 h) day 1 REF: Eisenhauer et al. J Clin Oncol 1994; 2654-2666 PREMEDICATIONS 1. Dexamethasone 20 mg IV 30 minutes before chemotherapy OR Dexamethasone 20 mg PO 6 and 12 hours prior to chemotherapy 2. Cimetidine 300 mg IV 30 minutes before chemotherapy 3. Diphenhydramine 25-50 mg IV 30 minutes before chemotherapy 4. Compazine 10 mg PO/IV 30 minutes before chemotherapy OTHER MEDICATIONS 1. Dexamethasone 4 mg PO BID for 6 doses after (for myalgias) Repeat every 21 days
Topotecan	Topotecan 1.5 mg/M² IV (over 30 min) days 1-5 REF: McGuire et al. J Clin Oncol 2000; 18:1062-1067 PREMEDICATIONS 1. Kytril 1 mg IV/PO 30 minutes before and 12 hours after chemotherapy on days 1-5 2. Dexamethasone 10 mg IV 30 minutes before chemotherapy on days 1-5 Repeat every 21 days Topotecan—hold for ANC < 1500 or platelets < 100,000; decrease dose by 0.25 mg/M²/d for prior episode of severe neutropenia or administer G-CSF starting on day 6

Trophoblastic Disease

Agent	Dosage			
LOW RISK DISEASE				
Dactinomycin	Dactinomycin	1.25 mg/M²	IV	day 1
	REF: Osathanondh et al. Cancer 1975; 36:863-866			
	PREMEDICATIONS 1. Kytril 1 mg PO/IV 30 minutes before and 12 hours after chemotherapy 2. Dexamethasone 20 mg IV 30 minutes before chemotherapy			
	Repeat every 14 days; treat for 1 to 2 cycles beyond negative HCG titers			
	Dactinomycin—vesicant–watch for extravasation			
Methotrexate	Methotrexate	40 mg/M²	IM	weekly
	REF: Gleeson et al. Eur J Gynaecol Oncol 1993; 14:461-465			
	Treat for 2 courses beyond negative HCG titers			
	Methotrexate—use 75% dose for CrCl < 50; 50% dose if CrCl < 25; do not give if patient has an effusion ("reservoir effect")			
INTERMEDIATE/HIGH RISK DISEASE				
EMA-CO etoposide (VP-16)/ dactinomycin/ methotrexate/ vincristine/ cyclophosphamide	Etoposide	100 mg/M²	IV	days 1, 2
	Dactinomycin	0.5 mg	IV	days 1, 2
	Methotrexate	100 mg/M²	IV	day 1
	—followed by			
	Methotrexate	200 mg/M²	CIV (over 12 h)	day 1
	Folic Acid	15 mg	PO/IM BID for 4 doses, starting 24 h after first methotrexate dose	
	Vincristine	0.8 mg/M²	IV	day 8
	Cyclophosphamide	600 mg/M²	IV	day 8
	–patients with pulmonary metastases receive intrathecal methotrexate every 2 weeks with cycles of CO			
	REF: Bower et al. J Clin Oncol 1997; 15:2636-2643			
	PREMEDICATIONS 1. Kytril 1 mg PO/IV 30 minutes before and 12 hours after chemotherapy on days 1, 2, and 8 2. Dexamethasone 20 mg IV 30 minutes before chemotherapy on days 1, 2, and 8			
	Repeat every 14 days			

Continued

Agent	Dosage			
	Dactinomycin—vesicant–watch for extravasation Methotrexate—use 75% dose for CrCl < 50; 50% dose if CrCl < 25; do not give if patient has an effusion ("reservoir effect") Vincristine—vesicant–avoid extravasation; cumulative neurotoxicity–may produce severe constipation; maximum 2 mg per administration			
EP/EMA etoposide (VP-16)/ cisplatin/ dactinomycin/ methotrexate	Etoposide	150 mg/M^2	IV	day 1
	Cisplatin	75 mg/M^2	IV (over 12 h)	day 1
	Etoposide	100 mg/M^2	IV	day 8
	Methotrexate	300 mg/M^2	IV (over 12 h)	day 8
	Dactinomycin	0.5 mg	IV	day 8
	Folinic Acid	15 mg	PO/IM BID –for 4 doses, starting 24 h after MTX	days 9, 10
	REF: Newlands et al. J Clin Oncol 2000; 18:854-859 PREMEDICATIONS 1. Kytril 1 mg PO/IV 30 minutes before and 12 hours after chemotherapy on days 1 and 8 2. Dexamethasone 20 mg IV 30 minutes before chemotherapy on days 1 and 8 OTHER MEDICATIONS 1. Give cisplatin delayed-emesis prophylaxis Repeat every 14 days Cisplatin—vigorous hydration is required; can be nephrotoxic and ototoxic; can cause peripheral neuropathy; hold or reduce for creatinine > 1.5 Dactinomycin—vesicant–watch for extravasation Methotrexate—use 75% dose for CrCl < 50; 50% dose if CrCl < 25; do not give if patient has an effusion ("reservoir effect")			
PVB cisplatin/ vinblastine/ bleomycin	Cisplatin	20 mg/M^2	IV	days 1-5
	Vinblastine	0.15 mg/kg	IV	days 1, 2
	Bleomycin	30 units	IV	days 2,9,16
	REF: Hainsworth et al. Cancer Treat Rep 1983; 67:393-395 PREMEDICATIONS 1. Kytril 1 mg PO/IV 30 minutes before and 12 hours after chemotherapy on days 1-5 2. Dexamethasone 10 mg IV 30 minutes before chemotherapy on days 1-5			

Continued

Agent	Dosage
	OTHER MEDICATIONS 1. Give cisplatin delayed-emesis prophylaxis Repeat every 21 days Cisplatin—vigorous hydration is required; can be nephrotoxic and ototoxic; can cause peripheral neuropathy; hold or reduce for creatinine > 1.5 Bleomycin—give test dose of 1-2 units because of possible acute pulmonary, anaphylactoid, or severe febrile reactions; must dose adjust for renal insufficiency; total lifetime dose should not exceed 400 units; avoid high FiO_2 as it can exacerbate pulmonary toxicity Vinblastine—use 50% of dose for bilirubin > 3.0; vesicant–avoid extravasation; watch for neurotoxicity
Paclitaxel	Paclitaxel 250 mg/M^2 CIV (over 24 h) day 1 REF: Termrungruanglert et al. Anticancer Drugs 1996; 7:503-506 PREMEDICATIONS 1. Dexamethasone 20 mg IV 30 minutes before chemotherapy OR Dexamethasone 20 mg PO 6 and 12 hours prior to chemotherapy 2. Cimetidine 300 mg IV 30 minutes before chemotherapy 3. Diphenhydramine 25-50 mg IV 30 minutes before chemotherapy 4. Compazine 10 mg PO/IV 30 minutes before chemotherapy OTHER MEDICATIONS 1. Dexamethasone 4 mg PO BID for 6 doses after (for myalgias) 2. Requires G-CSF support Repeat every 21 days

Chapter 8
Head and Neck

Head and Neck

Agent	Dosage			
CABO cisplatin/ methotrexate/ bleomycin/ vincristine	Cisplatin	50 mg/M²	IV	day 4
	Methotrexate	40 mg/M²	IV	days 1,15
	Bleomycin	10 units	IV	days 1,8,15
	Vincristine	2 mg	IV	days 1,8,15

–after 3 courses, methotrexate is given as weekly maintenance

–vincristine is discontinued after 6 doses

REF: Clavel et al. Cancer 1987; 60:1173-1177

PREMEDICATIONS

1. Kytril 1 mg PO/IV 30 minutes before and 12 hours after chemotherapy on day 4
2. Dexamethasone 20 mg IV 30 minutes before chemotherapy on day 4
3. Compazine 10 mg PO/IV 30 minutes before chemotherapy on days 1, 8, and 15

OTHER MEDICATIONS

Repeat every 21 days

1. Give cisplatin delayed-emesis prophylaxis

Cisplatin—vigorous hydration is required; can be nephrotoxic and ototoxic; can cause peripheral neuropathy; hold or reduce for creatinine > 1.5

Bleomycin—give test dose of 1-2 units because of possible acute pulmonary, anaphylactoid, or severe febrile reactions; must dose adjust for renal insufficiency; total lifetime dose should not exceed 400 units; avoid high FiO_2 as it can exacerbate pulmonary toxicity

Methotrexate—use 75% dose for CrCl < 50; 50% dose if CrCl < 25; do not give if patient has an effusion ("reservoir effect")

Vincristine—vesicant–avoid extravasation; cumulative neurotoxicity—may produce severe constipation; maximum 2 mg per administration

Agent	Dosage			
Carboplatin/ paclitaxel	Paclitaxel	200 mg/M²	IV (over 3 h)	day 1
	Carboplatin	AUC 7	IV	day 1

REF: Fountzilas et al. Ann Oncol 1997; 8:451-455

Continued

Agent	Dosage
	PREMEDICATIONS 1. Dexamethasone 20 mg IV 30 minutes before chemotherapy 2. Diphenhydramine 50 mg IV 30 minutes before chemotherapy 3. Cimetidine 300 mg IV 30 minutes before chemotherapy 4. Kytril 1 mg PO/IV 30 minutes before and 12 hours after chemotherapy OTHER MEDICATIONS 1. Dexamethasone 4 mg PO BID for 6 doses after paclitaxel (for myalgias) 2. G-CSF 5 mcg/kg/d SQ is given days 2-12 3. Give cisplatin delayed emesis prophylaxis Repeat every 21 days
CF cisplatin/ fluorouracil (5-FU)	Cisplatin 100 mg/M² IV day 1 5-FU 1000 mg/M²/d CIV days 1-4 REF: Kish et al. Cancer 1984; 53:1819-1824 PREMEDICATIONS 1. Kytril 1 mg PO/IV 30 minutes before and 12 hours after chemotherapy on day 1 2. Dexamethasone 20 mg IV 30 minutes before chemotherapy on day 1 OTHER MEDICATIONS 1. Give cisplatin delayed-emesis prophylaxis Repeat every 28 days Cisplatin—vigorous hydration is required; can be nephrotoxic and ototoxic; can cause peripheral neuropathy; hold or reduce for creatinine > 1.5
PF cisplatin fluorouracil (5-FU)/ XRT larynx preservation	Cisplatin 100 mg/M² IV day 1 5-FU 1000 mg/M²/d CIV days 1-5 –followed by XRT to 6600-7600 cGy REF: Veterans Affairs Laryngeal Cancer Study Group. NEJM 1991; 324:1685-1690 PREMEDICATIONS 1. Kytril 1 mg PO/IV 30 minutes before and 12 hours after chemotherapy on day 1 2. Dexamethasone 20 mg IV 30 minutes before chemotherapy on day 1

Continued

Agent	Dosage			
	OTHER MEDICATIONS 1. Give cisplatin delayed-emesis prophylaxis Repeat every 28 days Cisplatin—vigorous hydration is required; can be nephrotoxic and ototoxic; can cause peripheral neuropathy; hold or reduce for creatinine > 1.5			
PT cisplatin/ paclitaxel	Paclitaxel	200 mg/M²	IV (over 3 h)	day 1
	Cisplatin	75 mg/M²	IV	day 1
	REF: Hitt et al. Semin Oncol 1995; 22:50-54 PREMEDICATIONS 1. Dexamethasone 20 mg IV 30 minutes before chemotherapy 2. Diphenhydramine 50 mg IV 30 minutes before chemotherapy 3. Cimetidine 300 mg IV 30 minutes before chemotherapy 4. Kytril 1 mg PO/IV 30 minutes before and 12 hours after chemotherapy OTHER MEDICATIONS 1. Dexamethasone 4 mg PO BID for 6 doses after paclitaxel (for myalgias) 2. Give cisplatin delayed-emesis prophylaxis 3. G-CSF 5 mcg/kg/d SQ is given days 4-12 Repeat every 21-28 days Cisplatin—vigorous hydration is required; can be nephrotoxic and ototoxic; can cause peripheral neuropathy; hold or reduce for creatinine > 1.5			
TIP paclitaxel/ ifosfamide/ cisplatin	Paclitaxel	175 mg/M²	IV (over 3 h)	day 1
	Ifosfamide	1000 mg/M²	IV (over 2 h)	days 1-3
	Mesna	600 mg/M²	IV	days 1-3
	Cisplatin	60 mg/M²	IV	day 1
	REF: Shin et al. J Clin Oncol 1998; 16:1325-1330 PREMEDICATIONS 1. Dexamethasone 20 mg IV 30 minutes before chemotherapy on days 1-3 2. Diphenhydramine 50 mg IV 30 minutes before chemotherapy on day 1 3. Cimetidine 300 mg IV 30 minutes before chemotherapy on day 1 4. Kytril 1 mg PO/IV 30 minutes before and 12 hours after chemotherapy on days 1-3			

Continued

Agent	Dosage
	OTHER MEDICATIONS 1. Dexamethasone 4 mg PO BID for 6 doses after paclitaxel (for myalgias) 2. Give cisplatin delayed-emesis prophylaxis Repeat every 21-28 days Cisplatin—vigorous hydration is required; can be nephrotoxic and ototoxic; can cause peripheral neuropathy; hold or reduce for creatinine > 1.5 Ifosfamide—adequate hydration is necessary to prevent nephrotoxicity
VP vinorelbine/ cisplatin	Vinorelbine 25 mg/M² IV days 1,8 Cisplatin 80 mg/M² IV day 1 REF: Gebbia et al. Am J Clin Oncol 1995; 18:293-296 PREMEDICATIONS 1. Kytril 1 mg PO/IV 30 minutes before and 12 hours after chemotherapy on day 1 2. Dexamethasone 20 mg IV 30 minutes before chemotherapy on day 1 OTHER MEDICATIONS 1. Dexamethasone 4 mg PO BID for 6 doses after Paclitaxel (for myalgias) 2. Give cisplatin delayed-emesis prophylaxis Repeat every 21 days Cisplatin—vigorous hydration is required; can be nephrotoxic and ototoxic; can cause peripheral neuropathy; hold or reduce for creatinine > 1.5 Vinorelbine—vesicant; avoid extravasation; can cause peripheral neuropathy
Docetaxel	Docetaxel 100 mg/M² IV (over 1 h) day 1 REF: Dreyfuss et al. J Clin Oncol 1996; 14:1672-1678 PREMEDICATIONS 1. Dexamethasone 20 mg IV 30 minutes before chemotherapy 2. Cimetidine 300 mg IV 30 minutes before chemotherapy 3. Diphenhydramine 25-50 mg IV 30 minutes before chemotherapy 4. Compazine 10 mg PO/IV 30 minutes before chemotherapy

Continued

Agent	Dosage
	OTHER MEDICATIONS 1. Dexamethasone 8 mg PO BID for 8 doses—start day prior to chemo (decreases lower extremity edema) Repeat every 21 days
Methotrexate	Methotrexate 40 mg/M² IV day 1 REF: Forastiere et al. J Clin Oncol 1992; 10:1245-1251 PREMEDICATIONS 1. Compazine 10 mg PO/IV 30 minutes before chemotherapy Repeat every 7 days Methotrexate—use 75% dose for CrCl < 50; 50% dose if CrCl < 25; do not give if patient has an effusion ("reservoir effect")
Paclitaxel	Paclitaxel 250 mg/M² CIV (over 24 h) day 1 REF: Forastiere et al. Cancer 1998; 82:2270-2274 PREMEDICATIONS 1. Dexamethasone 20 mg IV 30 minutes before chemotherapy OR Dexamethasone 20 mg PO 6 and 12 hours prior to chemotherapy 2. Cimetidine 300 mg IV 30 minutes before chemotherapy 3. Diphenhydramine 25-50 mg IV 30 minutes before chemotherapy 4. Compazine 10 mg PO/IV 30 minutes before chemotherapy OTHER MEDICATIONS 1. Dexamethasone 4 mg PO BID for 6 doses after (for myalgias) 2. Requires G-CSF support Repeat every 21 days

Chapter 9
Hematologic Malignancies

- Acute Lymphocytic Leukemia
- Acute Myelogenous Leukemia
- Chronic Lymphocytic Leukemia
- Chronic Myelogenous Leukemia
- Hairy Cell Leukemia
- Hodgkin's Disease
- Multiple Myeloma
    - Waldenstrom's Macroglobulinemia
- Myelodysplastic Syndrome
- Non-Hodgkin's Lymphoma

Hematologic Malignancies

Acute Lymphocytic Leukemia

Hoelzer Regimen (BFM)

INDUCTION—PHASE I				
	Vincristine	2 mg	IV	days 1,8,15,22
	Daunorubicin	25 mg/M²	IV	days 1,8,15,22
	Prednisone	60 mg/M²	PO	days 1-28
	L-asparaginase	5,000 units/M²	IV	days 1-14
INDUCTION—PHASE II				
	Cyclophosphamide –maximum dose 1000 mg	650 mg/M²	IV	days 29,43,57
	Ara-C	75 mg/M²	IV	days 31-34,38-41, 45-48, 52-55
	6-Mercaptopurine	60 mg/M²	PO	days 29-57
CNS PROPHYLAXIS—weeks 5 through 8				
	Methotrexate –maximum dose is 15 mg	10 mg/M²	IT	days 31,38,45,52
	Cranial radiotherapy	1800-2400 cGy		given with phase II induction
CONSOLIDATION—PHASE I—begins week 20				
	Vincristine	2 mg	IV	days 1,8,15,22
	Doxorubicin	25 mg/M²	IV	days 1,8,15,22
	Dexamethasone	10 mg/M²	PO	days 1-28
CONSOLIDATION – PHASE II				
	Cyclophosphamide – maximum dose is 1000 mg	650 mg/M²	IV	day 29
	Ara-C	75 mg/M²	IV	days 31-34,38-41
	6-Thioguanine	60 mg/M²	PO	days 29-42
MAINTENANCE				
	6-Mercaptopurine	60 mg/M²	PO	daily weeks 10-18,29-130
	Methotrexate	20 mg/M²	PO/IV	weekly weeks 10-18,29-130
	REF: Hoelzer et al. Blood 1988; 71:123-131			
	PREMEDICATIONS 1. Kytril 1 mg PO/IV 30 minutes before and 12 hours after: daunorubicin, doxorubicin, and cyclophosphamide 2. Compazine 10 mg PO/IV 30 minutes before: cytarabine and L-asparaginase			

Continued

Brain Cancer | Breast Cancer | Carcinoma of Unknown Primary | Endocrine Cancer | Gastrointestinal Cancer | Genitourinary Cancer | Gynecologic Cancer | Head and Neck Cancer | Hematologic Malignancies

Anthracyclines—monitor cumulative dose for possible cardiac toxicity; vesicant–avoid extravasation

Methotrexate—use 75% dose for CrCl < 50; 50% dose if CrCl < 25; do not give if patient has an effusion ("reservoir effect")

Vincristine—vesicant–avoid extravasation; cumulative neurotoxicity—may produce severe constipation; maximum 2 mg per administration

6-Mercaptopurine—reduce dose by 75% when used in conjunction with allopurinol

L-asparaginase—be prepared to treat anaphylaxis at each administration; giving with or immediately before Vincristine may increase Vincristine toxicity

Hyper CVAD Regimen

HYPER CVAD ALTERNATING WITH HIGH DOSE METHOTREXATE/ARA-C

–alternate above for a total of 8 cycles

–subsequent cycles given when WBC recovers to > 3.0 and platelet count is > 60,000

HYPER CVAD—cycles 1, 3, 5, and 7

Cyclophosphamide	300 mg/M²	IV Q12H (over 3 h)	days 1-3
Mesna	600 mg/M²/d	CIV	days 1-3

–start at same time as cyclophosphamide and finish 6 hours after completion of cyclophosphamide

Vincristine	2 mg	IV	days 4,11
Doxorubicin	50 mg/M²	IV	day 4
Dexamethasone	40 mg	PO	days 1-4, 11-14

PREMEDICATIONS

1. Kytril 1 mg PO/IV 30 minutes before and 12 hours after chemotherapy on days 1-4

OTHER MEDICATIONS

1. Levofloxacin 500 mg PO QD
2. Fluconazole 200 mg PO QD
3. Valacyclovir 500 mg PO QD
4. Neupogen 10 mcg/kg/d SQ divided BID starting day 5

HIGH DOSE METHOTREXATE AND CYTARABINE (ARA-C)—cycles 2, 4, 6, 8

Methotrexate	200 mg/M²	IV (over 2 h)	day 1
–followed by			
Methotrexate	800 mg/M²	CIV (over 24 h)	day 1

Continued

Leucovorin	15 mg	PO Q6H for 8 doses	

–increase Leucovorin to 50 mg PO Q6H if methotrexate level is:

> 20 μmol/L at end of infusion

> 1 μmol/L 24 hr later

> 0.1 μmol/L 48 hr after the end of the methotrexate infusion

- continue until methotrexate level is < 0.1 μmol/L

Ara-C	3 gm/M²	IV over 2 hr Q12H for 4 doses	days 2-3
Methylprednisolone	50 mg	IV BID	days 1-3

PREMEDICATIONS
1. Kytril 1 mg PO/IV 30 minutes before and 12 hours after chemotherapy on days 1-3

OTHER MEDICATIONS
1. Levofloxacin 500 mg PO QD
2. Fluconazole 200 mg PO QD
3. Valacyclovir 500 mg PO QD
4. Neupogen 10 mcg/kg/d SQ divided BID starting day 5
5. Dexamethasone eye drops 2 drops each eye Q3H during and for 48-72 hours after completion of cytarabine

CNS TREATMENT/PROPHYLAXIS

High Risk—LDH > 600 and/or high proliferative index; mature B-cell ALL

Low Risk—neither of above

Methotrexate	12 mg	IT	day 2
Ara-C	100 mg	IT	day 8

Known CNS disease—IT therapy twice weekly until CNS negative, then per prophylaxis protocol

High risk—above is repeated for each of the 8 cycles of chemotherapy

Low risk—above is repeated only during the first 2 cycles of chemotherapy

Unknown risk—above is repeated during the first 4 cycles of chemotherapy

MAINTENANCE PHASE

A. Mature B-cell ALL—no maintenance
B. Ph+ ALL—allogeneic transplant if donor available; otherwise, IFN and Ara-C as below

–therapy is continued for 2 years

Continued

Brain Cancer | Breast Cancer | Carcinoma of Unknown Primary | Endocrine Cancer | Gastrointestinal Cancer | Genitourinary Cancer | Gynecologic Cancer | Head and Neck Cancer | Hematologic Malignancies

Interferon alfa	5 MIU/M^2	SQ	QD
Ara-C	10 mg	SQ	QD

C. All other patients

–therapy is continued for 2 years

6-Mercaptopurine	50 mg	PO TID	QD
Methotrexate	20 mg/M^2	PO	weekly
Vincristine	2 mg	IV	monthly
Prednisone	200 mg	PO	days 1-5 monthly

OTHER MEDICATIONS

1. Trimethoprim/sulfamethoxazole DS 1 tab PO BID each weekend
2. Valacyclovir 500 mg PO QD or TIW

–above medications are continued for first 6 months of maintenance phase

REF: Kantarjian et al. J Clin Oncol 2000; 18:547-561

Methotrexate—25% dose reduction for creatinine 1.5-2 and 50% reduction for creatinine > 2; do not give if patient has an effusion ("reservoir effect")

Vincristine—vesicant–avoid extravasation; cumulative neurotoxicity—may produce severe constipation; maximum 2 mg per administration; dose reduced to 1 mg for bilirubin > 2

6-Mercaptopurine—reduce dose by 75% when used in conjunction with allopurinol

Doxorubicin—monitor cumulative dose for cardiac toxicity (not to exceed 550 mg/M^2 or 450 mg/M^2 with prior chest Radiotherapy); vesicant—avoid extravasation; dose reduced by 25% if bilirubin 2-3, 50% if bilirubin 3-4, and 75% if bilirubin > 4

Ara-C—high doses can cause CNS toxicity (cerebellar dysfunction); neurotoxicity increases as infusion time increases; dose reduced to 1 gm/M^2 if age > 60, creatinine > 2, or if Methotrexate level at end of infusion is > 20 μmol/L

Larson Regimen

COURSE I: INDUCTION (4 WEEK)

WEEKS 1-4	Cyclophosphamide	1200 mg/M^2	IV	day 1
	Daunorubicin	45 mg/M^2	IV	days 1-3
	Vincristine	2 mg	IV	days 1,8, 15,22
	Prednisone	60 mg/M^2	PO	days 1-21
	L-asparaginase	6000 IU/M^2	SQ	days 5,8, 11,15,18,22

Continued

	–for patients > age 60, modify doses as follows:			
	Cyclophosphamide	800 mg/M²		on day 1
	Daunorubicin	30 mg/M²		on days 1-3
	Prednisone	60 mg/M²		on days 1-7
COURSE II: EARLY INTENSIFICATION (4 WEEK; REPEAT ONCE)				
WEEKS 5-12	Methotrexate	15 mg	IT	day 1
	Cyclophosphamide	1000 mg/M²	IV	day 1
	6-Mercaptopurine	60 mg/M²	PO	days 1-14
	Ara-C	75 mg/M²	SQ	days 1-4, 8-11
	Vincristine	2 mg	IV	days 15,22
	L-asparaginase	6,000 IU/M²	SQ	days 15,18, 22,25
COURSE III: CNS PROPHYLAXIS AND INTERIM MAINTENANCE (12 WEEK)				
WEEKS 13-25	Cranial Radiotherapy	2400 cGy		over days 1-12
	Methotrexate	15 mg	IT	days 1,8,15, 22,29
	6-Mercaptopurine	60 mg/M²	PO	days 1-70
	Methotrexate	20 mg/M²	PO	days 36,43, 50,57,64
COURSE IV: LATE INTENSIFICATION (8 WEEK)				
WEEKS 26-33	Doxorubicin	30 mg/M²	IV	days 1,8,15
	Vincristine	2 mg	IV	days 1,8,15
	Dexamethasone	10 mg/M²	PO	days 1-14
	Cyclophosphamide	1000 mg/M²	IV	day 29
	6-Thioguanine	60 mg/M²	PO	days 29-42
	Ara-C	75 mg/M²	SQ	days 29,32, 36-39
COURSE V: PROLONGED MAINTENANCE				
UNTIL 24 MONTHS FROM DIAGNOSIS				
	Vincristine	2 mg	IV	day 1 every 4 wks
	Prednisone	60 mg/M²	PO	days 1-5 every 4 wks
	Methotrexate	20 mg/M²	PO	days 1,8,15,22 every 4 wks
	6-Mercaptopurine	80 mg/M²	PO	days 1-28 every 4 wks
	REF: Larson et al. Blood 1995; 85:2025-2037			

PREMEDICATIONS

1. Kytril 1 mg PO/IV 30 minutes before and 12 hours after: daunorubicin, doxorubicin, and cyclophosphamide
2. Compazine 10 mg PO/IV 30 minutes before: cytarabine and L-asparaginase

Continued

Brain Cancer
Breast Cancer
Carcinoma of Unknown Primary
Endocrine Cancer
Gastrointestinal Cancer
Genitourinary Cancer
Gynecologic Cancer
Head and Neck Cancer
Hematologic Malignancies

Anthracyclines—monitor cumulative dose for possible cardiac toxicity; vesicant–avoid extravasation

Methotrexate—use 75% dose for CrCl < 50; 50% dose if CrCl < 25; do not give if patient has an effusion ("reservoir effect")

Vincristine—vesicant–avoid extravasation; cumulative neurotoxicity–may produce severe constipation; maximum 2 mg per administration

6-Mercaptopurine—reduce dose by 75% when used in conjunction with allopurinol

L-asparaginase—be prepared to treat anaphylaxis at each administration; giving with or immediately before Vincristine may increase Vincristine toxicity

Linker Regimen

INDUCTION

Drug	Dose	Route	Schedule
Daunorubicin	50 mg/M²	IV	days 1-3
Vincristine	2 mg	IV	days 1,8,15,22
Prednisone	60 mg/M²	PO divided TID	days 1-28
L-asparaginase	6,000 IU/M²	IM	days 17-28

–if day 14 bone marrow shows residual leukemia

Drug	Dose	Route	Schedule
Daunorubicin	50 mg/M²	IV	day 15

–if day 28 bone marrow shows residual leukemia

Drug	Dose	Route	Schedule
Daunorubicin	50 mg/M²	IV	days 29,30
Vincristine	2 mg	IV	days 29,36
Prednisone	60 mg/M²	PO divided TID	days 29-42
L-asparaginase	6,000 IU/M²	IM	days 29-35

CNS PROPHYLAXIS

–initiate within 1 week of achieving complete remission

Treatment	Dose	Route	Schedule
Cranial XRT	1800 cGy		in 10 fractions
Methotrexate	12 mg	IT	weekly X 6

–if CNS is positive at time of diagnosis

–begin weekly intrathecal MTX during induction

–MTX 12 mg IT weekly X 10

–Cranial XRT to 2800 cGy

CONSOLIDATION—TREATMENT A—CYCLES 1, 3, 5, 7

Drug	Dose	Route	Schedule
Daunorubicin	50 mg/M²	IV	days 1,2
Vincristine	2 mg	IV	days 1,8
Prednisone	60 mg/M²	PO divided TID	days 1-14
L-asparaginase	12,000 IU	IM	days 2,4,7, 9,11,14

Continued

CONSOLIDATION—TREATMENT B—CYCLES 2, 4, 6, 8				
	Teniposide	165 mg/M²	IV	days 1,4,8,11
	Ara-C	300 mg/M²	IV	days 1,4,8,11
CONSOLIDATION—TREATMENT C—COURSE 9				
	Methotrexate	690 mg/M²	IV (over 42 h)	day 1
	Leucovorin	15 mg/M²	IV Q6H for 12 doses—start at hour 42	
MAINTENANCE THERAPY				
	–continued for 30 months of CR			
	Methotrexate	20 mg/M²	PO	weekly
	6-MP	75 mg/M²	PO	daily
	REF: Linker et al. Blood 1991; 78:2814-2822			

PREMEDICATIONS
1. Kytril 1 mg PO/IV 30 minutes before and 12 hours after daunorubicin
2. Compazine 10 mg PO/IV 30 minutes before Ara-C, L-asparaginase, and teniposide

Daunorubicin—monitor cumulative dose for possible cardiac toxicity; vesicant–avoid extravasation

Methotrexate—use 75% dose for CrCl < 50; 50% dose if CrCl < 25; do not give if patient has an effusion ("reservoir effect")

Vincristine—vesicant–avoid extravasation; cumulative neurotoxicity—may produce severe constipation; maximum 2 mg per administration

6-Mercaptopurine—reduce dose by 75% when used in conjunction with allopurinol

L-asparaginase—be prepared to treat anaphylaxis at each administration; giving with or immediately before Vincristine may increase Vincristine toxicity

Acute Myelogenous Leukemia

INDUCTION CHEMOTHERAPY

Agent	Dosage
7+3 cytarabine (ara-c)/ daunorubicin	Ara-C 100 mg/M²/d CIV days 1-7 Daunorubicin 45 mg/M² IV days 1-3 REF: Yates et al. Blood 1982; 60:454-462 PREMEDICATIONS 1. Kytril 1 mg PO/IV 30 minutes before and Q12 hours during chemotherapy on days 1-7 2. Dexamethasone 20 mg IV 30 minutes before chemotherapy on days 1-3 Daunorubicin—monitor cumulative dose for possible cardiac toxicity; vesicant—avoid extravasation CONSOLIDATION—repeat the above drugs for 5 and 2 days respectively
7+3+7 cytarabine (ara-c)/ daunorubicin/ etoposide (VP-16)	Ara-C 100 mg/M²/d CIV days 1-7 Daunorubicin 50 mg/M² IV days 1-3 VP-16 75 mg/M² IV (over 1 h) days 1-7 REF: Bishop et al. Blood 1990; 75:27-32 PREMEDICATIONS 1. Kytril 1 mg PO/IV 30 minutes before and Q12 hours during chemotherapy on days 1-7 2. Dexamethasone 20 mg IV 30 minutes before chemotherapy on days 1-3 Daunorubicin—monitor cumulative dose for possible cardiac toxicity; vesicant—avoid extravasation CONSOLIDATION—repeat the cytarabine for 5 days and the daunorubicin for 2 days (and optional 5 days of etoposide)
Idarubicin/ cytarabine (ara-c)	Ara-C 100 mg/M²/d CIV days 1-7 Idarubicin 13 mg/M² IV days 1-3 REF: Wiernick et al. Blood 1992; 79:313-319 PREMEDICATIONS 1. Kytril 1 mg PO/IV 30 minutes before and Q12 hours during chemotherapy on days 1-7 2. Dexamethasone 20 mg IV 30 minutes before chemotherapy on days 1-3

Continued

Agent	Dosage
	Idarubicin—monitor cumulative dose for possible cardiac toxicity; vesicant—avoid extravasation CONSOLIDATION—repeat the above drugs for 5 and 2 days respectively
Mitoxantrone/ cytarabine (ara-c)	Ara-C 100 mg/M²/d CIV days 1-7 Mitoxantrone 12 mg/M² IV days 1-3 REF: Arlin et al. Leukemia 1990; 4:177-183 PREMEDICATIONS 1. Kytril 1 mg PO/IV 30 minutes before and Q12 hours during chemotherapy on days 1-5 2. Dexamethasone 20 mg IV 30 minutes before chemotherapy on days 1 and 2 Mitoxantrone—watch cumulative dose—do not exceed 140 mg/M²; possible cardiac toxicity CONSOLIDATION—repeat the above drugs for 5 and 2 days respectively
TAD 9 daunorubicin/ cytarabine (ara-c)/ 6-thioguanine (6-TG)	Ara-C 100 mg/M²/d CIV days 1-2 –followed by Ara-C 100 mg/M² IV Q12H (over 30 min) days 3-8 Daunorubicin 60 mg/M² IV days 3-5 6-TG 100 mg/M² PO Q12H days 3-9 REF: Buchner et al. J Clin Oncol 1985; 3:1583-1589 –there are several variations of the DAT/TAD regimen PREMEDICATIONS 1. Kytril 1 mg PO/IV 30 minutes before and Q12 hours during chemotherapy on days 1-8 2. Dexamethasone 20 mg IV 30 minutes before chemotherapy on days 1-5 Daunorubicin—monitor cumulative dose for possible cardiac toxicity; vesicant—avoid extravasation
CONSOLIDATION CHEMOTHERAPY	
HiDAC high-dose cytarabine (ara-c)	–has been used as consolidation chemotherapy or for recurrent disease Ara-C 3000 mg/M² IVQ12H (over 3 h) days 1,3,5 –note that this is given with an anthracycline, as in the above regimens REF: Mayer et al. NEJM 1994; 331:896-903

Continued

Brain Cancer | Breast Cancer | Carcinoma of Unknown Primary | Endocrine Cancer | Gastrointestinal Cancer | Genitourinary Cancer | Gynecologic Cancer | Head and Neck Cancer | Hematologic Malignancies

Agent	Dosage
	–there are several variations of the HiDAC regimen PREMEDICATIONS 1. Kytril 1 mg PO/IV 30 minutes before and 12 hours after chemotherapy on days 1, 3, and 5 2. Dexamethasone 20 mg IV 30 minutes before chemotherapy on days 1, 3, and 5 3. Dexamethasone eye drops 2 drops each eye Q3H during and for 48-72 hours after completion of cytarabine Repeat every 28 days (as consolidation) for 2 or 3 courses Ara-C—high doses can cause CNS toxicity (cerebellar dysfunction); neurotoxicity increases as infusion time increases
RELAPSED/REFRACTORY DISEASE	
HAM high-dose cytarabine (ara-c)/ mitoxantrone	Ara-C 3000 mg/M² IVQ12H(over 3 h) days 1-4 Mitoxantrone 10 mg/M² IV(over 30 min) days 2-5 or 6 REF: Hiddemann et al. Blood 1987; 69:744-749 PREMEDICATIONS 1. Kytril 1 mg PO/IV 30 minutes before and Q12 hours during chemotherapy on days 1-5 2. Dexamethasone 20 mg IV 30 minutes before chemotherapy on days 1-4 3. Dexamethasone eye drops 2 drops each eye Q3H during and for 48-72 hours after completion of cytarabine Ara-C—high doses can cause CNS toxicity (cerebellar dysfunction); neurotoxicity increases as infusion time increases Mitoxantrone–watch cumulative dose–do not exceed 140 mg/M²; possible cardiac toxicity
High-dose cytarabine (ara-c)/ fludarabine	Fludarabine 30 mg/M² IV(over 30 min) days 2-6 –followed 31/2 hours later by Ara-C 1000 mg/M² IV(over 2 h) days 1-6 REF: Estey et al. Leuk Lymphoma 1993; 9:343-350 PREMEDICATIONS 1. Kytril 1 mg PO/IV 30 minutes before and Q12 hours during chemotherapy on days 1-5 2. Dexamethasone 20 mg IV 30 minutes before chemotherapy on days 1-5 3. Dexamethasone eye drops 2 drops each eye Q3H during and for 48-72 hours after completion of cytarabine Ara-C—high doses can cause CNS toxicity (cerebellar dysfunction); neurotoxicity increases as infusion time increases

Continued

Agent	Dosage			
Mitoxantrone/ etoposide (VP-16)	INDUCTION			
	VP-16	100 mg/M²	IV	days 1-5
	Mitoxantrone	10 mg/M²	IV	days 1-5
	CONSOLIDATION			
	VP-16	75 mg/M²	IV	days 1-5
	Mitoxantrone	8 mg/M²	IV	days 1-5
	Ara-C	75 mg/M²	IV Q12H	days 1-5
	REF: Ho et al. J Clin Oncol 1988; 6:213-217			
	PREMEDICATIONS 1. Kytril 1 mg PO/IV 30 minutes before and Q12 hours during chemotherapy on days 1-5 2. Dexamethasone 20 mg IV 30 minutes before chemotherapy on days 1-5			
	Mitoxantrone—watch cumulative dose—do not exceed 140 mg/M²; possible cardiac toxicity			
Gemtuzumab zogamicin (Mylotarg)	–also called CMA-676			
	Mylotarg	9 mg/M2	IV	days 1,15
	REF: Sievers et al. Blood 1999; 94 (Suppl 1):abstract 3079			
	PREMEDICATIONS 1. Benadryl 25-50 mg PO/IV 30 minutes before 2. Tylenol 650 mg PO 30 minutes before Day 15 dose is given regardless of blood counts			
ACUTE PROMYELOCYTIC LEUKEMIA				
ATRA/ daunorubicin/ cytarabine (ara-c)	INDUCTION			
	ATRA	45 mg/M²	PO (divided BID)	daily until CR or 90 days
	Daunorubicin	60 mg/M²	IV	days 3-5
	Ara-C	200 mg/M²	IV	days 3-9
	CONSOLIDATION 1			
	Daunorubicin	60 mg/M²	IV	days 1-3
	Ara-C	200 mg/M²	IV	days 1-7
	CONSOLIDATION 2			
	Daunorubicin	45 mg/M²	IV	days 1-3
	Ara-C	1000 mg/M²	IV Q12H	days 1-4
	MAINTENANCE –continued to complete 2 years of therapy			
	ATRA	45 mg/M²	PO (divided BID for 15 days)	every 3 mos

Continued

Agent	Dosage
	6-MP 90 mg/M²/d PO daily Methotrexate 15 mg/M² PO weekly REF: Fenaux et al. Blood 1999; 94:1192-1200 PREMEDICATIONS 1. Kytril 1 mg PO/IV 30m minutes before and Q12 hours during daunorubicin and ara-c 2. Dexamethasone 20 mg IV 30 minutes before chemotherapy during daunorubicin and ara-c 6-Mercaptopurine—reduce dose by 75% when used in conjunction with allopurinol Daunorubicin—monitor cumulative dose for possible cardiac toxicity; vesicant—avoid extravasation Ara-C—high doses can cause CNS toxicity (cerebellar dysfunction); neurotoxicity increases as infusion time increases Methotrexate—use 75% dose for CrCl < 50; 50% dose if CrCl < 25; do not give if patient has an effusion ("reservoir effect")
AIDA **ATRA/idarubicin**	INDUCTION ATRA 45 mg/M² PO (divided BID) daily until CR or 90 days Idarubicin 12 mg/M² IV days 2,4,6,8 CONSOLIDATION 1 Idarubicin 5 mg/M² IV days 1-4 CONSOLIDATION 2 Mitoxantrone 10 mg/M² IV days 1-5 CONSOLIDATION 3 Idarubicin 12 mg/M² IV day 1 MAINTENANCE –continued to complete 2 years of therapy ATRA 45 mg/M² PO (divided BID for15 days) every 3 mos 6-MP 90 mg/M²/d PO daily Methotrexate 15 mg/M² IM weekly REF: Sanz et al. Blood 1999; 94:3015-3021

Continued

Agent	Dosage
	PREMEDICATIONS 1. Kytril 1 mg PO/IV 30m minutes before and Q12 hours during anthracycline therapy 2. Dexamethasone 20 mg IV 30 minutes before chemotherapy during anthracycline therapy 6-Mercaptopurine—reduce dose by 75% when used in conjunction with allopurinol Idarubicin—monitor cumulative dose for possible cardiac toxicity; vesicant – avoid extravasation Mitoxantrone—watch cumulative dose—do not exceed 140 mg/M^2; possible cardiac toxicity Methotrexate—use 75% dose for CrCl < 50; 50% dose if CrCl < 25; do not give if patient has an effusion ("reservoir effect")
Arsenic trioxide	Arsenic trioxide 0.1 mg/kg/d IV (over 1-2 h) days 1-28 REF: Westervelt et al. Blood 1999; 94 (Suppl 1):abstract 2268 Repeat every 42 days for a maximum of 3 cycles (or until cytogenetic remission, followed by 1 consolidation course)

Brain Cancer
Breast Cancer
Carcinoma of Unknown Primary
Endocrine Cancer
Gastrointestinal Cancer
Genitourinary Cancer
Gynecologic Cancer
Head and Neck Cancer
Hematologic Malignancies

Chronic Lymphocytic Leukemia

Agent	Dosage
COP cyclophos-phamide/ vincristine/ prednisone	Cyclophosphamide 400 mg/M² PO days 1-5 Vincristine 1.4 mg/M² IV day 1 Prednisone 80 mg PO days 1-5 REF: Raphael et al. J Clin Oncol 1991; 9:770-776 PREMEDICATIONS 1. Kytril 1 mg PO/IV 30 minutes before and 12 hours after chemotherapy on days 1-5 Repeat every 21 days Vincristine—vesicant–avoid extravasation; cumulative neurotoxicity; may produce severe constipation; maximum 2 mg per administration
FCR - cyclophos-phamide/ fludarabine/ rituximab	Cyclophosphamide 250 mg/M² IV days 1-3 Fludarabine 25 mg/M² IV days 1-3 Rituximab 375 mg/M² IV day 1 –for cycles 2-6, dose is increased to 500 mg/M² –infusion is started at 50 mg/hr (25 mg/hr in patients with circulating tumor cells) and slowly increased to a maximum of 400 mg/hr (300 mg/hr during initial infusion) REF: Keating et al. Proc Am Soc Clin Oncol 2000; abstract 2214 PREMEDICATIONS 1. Kytril 1 mg PO/IV 30 minutes before and 12 hours after chemotherapy 2. Tylenol 650 mg PO 30 minutes before rituximab 3. Benadryl 25 mg PO/IV 30 minutes before rituximab Trimethoprim-sulfamethoxazole DS BID for 2 days each week for patients who require any corticosteroids Repeat every 28 days
Chlorambucil daily	Chlorambucil 0.1 mg/kg PO QD REF: Dighiero et al. NEJM 1998; 338:1506-1514 Given daily –adjust dose based on CBC
Chlorambucil pulse	Chlorambucil 0.3 mg/kg PO days 1-5 Prednisone 40 mg/M² PO days 1-5 REF: Dighiero et al. NEJM 1998; 338:1506-1514

Continued

	Agent Dosage			
	OR			
	Chlorambucil	30 mg/M²	PO	day 1
	Prednisone	100 mg/M²	PO	days 1-5
	REF: Raphael et al. J Clin Oncol 1991; 9:770-776			
	Repeat every 28 days			
	–adjust dose based on CBC			
Cyclophos-phamide—oral	Cyclophosphamide	1-2 mg/kg	PO	daily
	REF: Huguley et al. Cancer Treat Rev 1977; 4:261-273			
	–there are multiple variations of this regimen			
	Cyclophosphamide—precautions against hemorrhagic cystitis			
Cyclophos-phamide–IV	Cyclophosphamide	20 mg/kg	IV	day 1
	REF: Huguley et al. Cancer Treat Rev 1977; 4:261-273			
	–there are multiple variations of this regimen			
	PREMEDICATIONS 1. Kytril 1 mg PO/IV 30 minutes before and 12 hours after chemotherapy 2. Dexamethasone 20 mg IV 30 minutes before chemotherapy			
	Repeat every 14-21 days			
	Cyclophosphamide—precautions against hemorrhagic cystitis			
Fludarabine	Consider prophylactic use of trimethoprin-sulfamethoxazole			
	Fludarabine	25 mg/M²	IV	days 1-5
	REF: Keating et al. J Clin Oncol 1991; 9:44-49			
	Repeat every 28 days			

Chronic Myelogenous Leukemia

Agent	Dosage			
Interferon-alfa 2a (IFN)/ cytarabine (Ara-C)	IFN	5 X 10^6 units/M^2	SQ	daily
	Ara-C	10 mg	SQ	daily
	REF: Kantarjian et al. J Clin Oncol 1999; 17:284-292			
	PREMEDICATIONS 1. Tylenol 650 mg PO before IFN 2. Compazine 10 mg PO before prn			
Busulfan	Busulfan	4-8 mg	PO	daily
	REF: Bolin et al. Cancer 1982; 50:1683-1686			
	Hold for WBC count < 20,000; resume for WBC > 50,000			
Hydroxyurea	Hydroxyurea	500-2000 mg	PO	daily
	REF: Bolin et al. Cancer 1982; 50:1683-1686			
Interferon-alfa 2a (IFN)	IFN	5 X 10^6 units/M^2	SQ	daily
	REF: Alimena et al. Blood 1988; 72:642-647			
	PREMEDICATIONS 1. Tylenol 650 mg PO before IFN prn			
	Interferon—adjust dose as tolerated to maintain WBC count 3000-5000			
Thiotepa	–this agent can be used for persistent thrombocythemia in CML patients who have adequate WBC count			
	Thiotepa	75 mg/M^2	IV	day 1
	REF: Rodriquez-Monge et al. Cancer 1997; 80:396-400			
	PREMEDICATIONS 1. Compazine 10 mg PO/IV 30 minutes before chemotherapy			
	Repeat every 14-21 days			

Hairy Cell Leukemia

Agent	Dosage
Cladribine (2-chlorode-oxyadenosine, 2-CdA)	2-CdA 0.1 mg/kg/d CIV days 1-7 REF: Piro et al. NEJM 1990; 322:1117-1121 Single 7 day infusion
Interferon alfa-2a (IFN)	IFN 2 MIU/M² SQ TIW for 1 year REF: Rai et al. Leukemia 1995; 9:1116-1120 PREMEDICATIONS 1. Acetaminophen as needed to alleviate fever or "flu-like" symptoms
Pentostatin (2-deoxycoformycin)	Pentostatin 4 mg/M² IV day 1 REF: Grever et al. J Clin Oncol 1995; 13:974-982 Repeat every 14 days for at least 3 months

Hodgkin's Disease

Agent	Dosage			
ABVD doxorubicin/ bleomycin/ vinblastine/ dacarbazine (DTIC)	Doxorubicin	25 mg/M^2	IV	days 1,15
	Bleomycin	10 mg/M^2	IV	days 1,15
	Vinblastine	6 mg/M^2	IV	days 1,15
	DTIC	375 mg/M^2	IV	days 1,15
	REF: Bonadonna et al. Cancer 1975; 36:252-259 PREMEDICATIONS 1. Kytril 1 mg PO/IV 30 minutes before and 12 hours after chemotherapy on days 1 and 15 2. Dexamethasone 20 mg IV 30 minutes before chemotherapy on days 1 and 15 Repeat every 28 days Doxorubicin—monitor cumulative dose for cardiac toxicity (not to exceed 550 mg/M^2 or 450 mg/M^2 with prior chest radiotherapy); vesicant—avoid extravasation; use 50% for bilirubin 1.5-3.0; use 25% for bilirubin > 3.0 Vinblastine—use 50% of dose for bilirubin > 3.0; vesicant–avoid extravasation; watch for neurotoxicity Bleomycin—give test dose of 1-2 units because of possible acute pulmonary, anaphylactoid, or severe febrile reactions; must dose adjust for renal insufficiency; total lifetime dose should not exceed 400 units; avoid high FiO_2 as it can exacerbate pulmonary toxicity Dacarbazine—vesicant—avoid extravasation			
ASHAP doxorubicin/ methyl-prednisolone/ cytarabine/ cisplatin	Doxorubicin	10 mg/M^2/d	CIV (for 96 h)	days 1-4
	Methylprednisolone	500 mg	IV (over 15 min)	days 1-4
	Cytarabine	1500 mg/M^2	IV (over 2 h)	day 5
	Cisplatin	25 mg/M^2/d	CIV (for 96 h)	days 1-4
	REF: Rodriguez et al. Blood 1999; 93:3632-3636 PREMEDICATIONS 1. Kytril 1 mg PO/IV Q12H for 10 doses, starting 30 minutes before chemotherapy on day 1 OTHER MEDICATIONS 1. Give cisplatin delayed-emesis prophylaxis Repeat every 21–28 days			

Continued

Agent	Dosage
	Doxorubicin—monitor cumulative dose for cardiac toxicity (not to exceed 550 mg/M² or 450 mg/M² with prior chest radiotherapy); vesicant—avoid extravasation; use 50% for bilirubin 1.5-3.0; use 25% for bilirubin > 3.0 Cisplatin—vigorous hydration is required; can be nephrotoxic and ototoxic; can cause peripheral neuropathy; hold or reduce for creatinine > 1.5 Cytarabine—high doses can cause CNS toxicity (cerebellar dysfunction); neurotoxicity increases as infusion time increases
BEACOPP cyclophos-phamide/ vincristine/ etoposide (VP-16)/ procarbazine/ prednisone/ doxorubicin/ bleomycin	Cyclophosphamide 650 mg/M² IV day 1 Vincristine 1.4 mg/M² IV day 1 VP-16 100 mg/M² IV days 1-3 Procarbazine 100 mg/M² PO days 1-7 Prednisone 40 mg/M² PO days 1-14 Doxorubicin 25 mg/M² IV day 1 Bleomycin 10 mg/M² IV day 8 REF: Tesch et al. Blood 1998; 92:4560-4567 PREMEDICATIONS 1. Kytril 1 mg PO/IV 30 minutes before and 12 hours after chemotherapy on day 1 2. Dexamethasone 20 mg IV 30 minutes before chemotherapy on day 1 3. Compazine 10 mg PO/IV 30 minutes before chemotherapy on day 8 Repeat every 28 days Vincristine—vesicant–avoid extravasation; cumulative neurotoxicity—may produce severe constipation; maximum 2 mg per administration Doxorubicin—monitor cumulative dose for cardiac toxicity (not to exceed 550 mg/M² or 450 mg/M² with prior chest radiotherapy); vesicant—avoid extravasation; use 50% for bilirubin 1.5-3.0; use 25% for bilirubin > 3.0 Bleomycin—give test dose of 1-2 units because of possible acute pulmonary, anaphylactoid, or severe febrile reactions; must dose adjust for renal insufficiency; total lifetime dose should not exceed 400 units; avoid high FiO_2 as it can exacerbate pulmonary toxicity

Brain Cancer | Breast Cancer | Carcinoma of Unknown Primary | Endocrine Cancer | Gastrointestinal Cancer | Genitourinary Cancer | Gynecologic Cancer | Head and Neck Cancer | Hematologic Malignancies

Agent	Dosage			
Dexa-BEAM dexamethasone/ carmustine (BCNU)/ etoposide (VP-16)/ cytarabine (Ara-C)/ melphalan	Dexamethasone	8 mg	PO Q8H	days 1-10
	BCNU	60 mg/M²	IV	day 2
	VP-16	75 mg/M²	IV	days 4-7
	Ara-C	100 mg/M²	IV Q12H	days 4-7
	Melphalan	20 mg/M²	IV	day 3
	REF: Pfreundschuh et al. J Clin Oncol 1994; 12:580-586			
	PREMEDICATIONS 1. Kytril 1 mg PO/IV 30 minutes before and 12 hours after chemotherapy on days 2 and 3 2. Compazine 10 mg PO/IV 30 minutes before chemotherapy on days 4-7			
	OTHER MEDICATIONS 1. Give non-cisplatin delayed emesis prophylaxis			
	Repeat every 28 days			
	Carmustine—maximum total dose is 1440 mg/M²; causes delayed myelosuppression			
Mini-BEAM carmustine (BCNU)/ etoposide (VP-16)/ cytarabine (Ara-C)/ melphalan	BCNU	60 mg/M²	IV	day 1
	VP-16	75 mg/M²	IV	days 2-5
	Ara-C	100 mg/M²	IV Q12H	days 2-5
	Melphalan	20 mg/M²	IV	day 6
	REF: Colwill et al. J Clin Oncol 1995; 13:396-402			
	PREMEDICATIONS 1. Kytril 1 mg PO/IV 30 minutes before and 12 hours after chemotherapy on days 1 and 6 2. Dexamethasone 20 mg IV 30 minutes before chemotherapy on days 1 and 6 3. Compazine 10 mg PO/IV 30 minutes before chemotherapy on days 2-5			
	OTHER MEDICATIONS 1. Give non-cisplatin delayed emesis prophylaxis			
	Repeat every 28-42 days			
	Carmustine—maximum total dose is 1440 mg/M²; causes delayed myelosuppression			
ChlVPP chlorambucil/ vinblastine/ procarbazine/ prednisone	Chlorambucil	6 mg/M²	PO	days 1-14
	Vinblastine	6 mg/M²	IV	days 1, 8
	–maximum dose is 10 mg			
	Procarbazine	100 mg/M²	PO	days 1-14
	Prednisone	40 mg/M²	PO	days 1-14

Continued

Agent	Dosage			
	REF: Selby et al. Br J Cancer 1990; 62:279-285 PREMEDICATIONS 1. Compazine 10 mg PO/IV 30 minutes before chemotherapy on days 1 and 8 Repeat every 28 days for 6 cycles Vinblastine—use 50% of dose for bilirubin > 3.0; vesicant–avoid extravasation; watch for neurotoxicity			
DHAP dexamethasone/ cytarabine (Ara-C)/cisplatin	Cisplatin	100 mg/M²	CIV (X 24 h)	day 1
	Ara-C	2000 mg/M²	IV Q12H X 2 doses, each over 3 h	day 2
	–start at completion of cisplatin infusion			
	Dexamethasone	40 mg	PO/IV	days 1-4
	REF: Velasquez et al. Blood 1988; 71:117-122 PREMEDICATIONS 1. Kytril 1 mg PO/IV 30 minutes before and 12 hours after chemo on days 1 and 2 OTHER MEDICATIONS 1. Give cisplatin delayed-emesis prophylaxis Repeat every 21-28 days Cisplatin—vigorous hydration is required; can be nephrotoxic and ototoxic; can cause peripheral neuropathy; hold or reduce for creatinine > 1.5 Cytarabine—high doses can cause CNS toxicity (cerebellar dysfunction); neurotoxicity increases as infusion time increases			
EVA etoposide (VP-16)/ vinblastine/ doxorubicin	VP-16	100 mg/M²	IV	days 1-3
	Vinblastine	6 mg/M²	IV	day 1
	Doxorubicin	50 mg/M²	IV	day 1
	REF: Canellos et al. J Clin Oncol 1995; 13:2005-2011 PREMEDICATIONS 1. Kytril 1 mg PO/IV 30 minutes before and 12 hours after chemotherapy on day 1 2. Dexamethasone 20 mg IV 30 minutes before chemotherapy on day 1 3. Compazine 10 mg PO/IV 30 minutes before etoposide on days 2 and 3 Repeat every 28 days			

Continued

Agent	Dosage
	Doxorubicin—monitor cumulative dose for cardiac toxicity (not to exceed 550 mg/M² or 450 mg/M² with prior chest radiotherapy); vesicant—avoid extravasation; use 50% for bilirubin 1.5-3.0; use 25% for bilirubin > 3.0 Vinblastine—use 50% of dose for bilirubin > 3.0; vesicant–avoid extravasation; watch for neurotoxicity
MOPP nitrogen mustard/ vincristine/ procarbazine/ prednisone	–this regimen is rarely utilized today, and is listed primarily for historical interest Nitrogen mustard 6 mg/M² IV days 1, 8 Vincristine 1.4 mg/M² IV days 1, 8 Procarbazine 100 mg/M² PO days 1-14 Prednisone 40 mg/M² PO days 1-14 REF: DeVita et al. Ann Intern Med 1970; 73:881-895 PREMEDICATIONS 1. Kytril 1 mg PO/IV 30 minutes before and 12 hours after chemotherapy on days 1 and 8 2. Dexamethasone 20 mg IV 30 minutes before chemotherapy on days 1 and 8 Repeat every 28 days Vincristine—vesicant–avoid extravasation; cumulative neurotoxicity—may produce severe constipation; maximum 2 mg per administration Nitrogen mustard—potent vesicant—avoid extravasation; decomposes rapidly after mixing; must not be mixed in same syringe with any other drug
MOPP-ABV Hybrid nitrogen mustard/ vincristine/ procarbazine/ prednisone/ doxorubicin/ bleomycin/ vinblastine	Nitrogen mustard 6 mg/M² IV day 1 Vincristine 1.4 mg/M² IV day 1 Procarbazine 100 mg/M² PO days 1-7 Prednisone 40 mg/M² PO days 1-14 Doxorubicin 35 mg/M² IV day 8 Bleomycin 10 mg/M² IV day 8 Vinblastine 6 mg/M² IV day 8 REF: Klimo et al. J Clin Oncol 1985; 3:1174-1182 PREMEDICATIONS 1. Kytril 1 mg PO/IV 30 minutes before and 12 hours after chemotherapy on days 1 and 8 2. Dexamethasone 20 mg IV 30 minutes before chemotherapy on days 1 and 8 Repeat every 28 days

Continued

Agent	Dosage
	Vincristine—vesicant–avoid extravasation; cumulative neurotoxicity—may produce severe constipation; maximum 2 mg per administration Nitrogen mustard—potent vesicant—avoid extravasation; decomposes rapidly after mixing; must not be mixed in same syringe with any other drug Doxorubicin—monitor cumulative dose for cardiac toxicity (not to exceed 550 mg/M² or 450 mg/M² with prior chest radiotherapy); vesicant – avoid extravasation; use 50% for bilirubin 1.5-3.0; use 25% for bilirubin > 3.0 Vinblastine—use 50% of dose for bilirubin > 3.0; vesicant—avoid extravasation; watch for neurotoxicity Bleomycin—give test dose of 1-2 units because of possible acute pulmonary, anaphylactoid, or severe febrile reactions; must dose adjust for renal insufficiency; total lifetime dose should not exceed 400 units; avoid high FiO_2 as it can exacerbate pulmonary toxicity
STANFORD V nitrogen mustard/ doxorubicin/ vinblastine/ vincristine/ bleomycin/ etoposide (VP-16)/ prednisone	Nitrogen mustard 6 mg/M² IV day 1 Doxorubicin 25 mg/M² IV days 1, 15 Vinblastine 6 mg/M² IV days 1, 15 Vincristine 1.4 mg/M² IV days 8, 22 Bleomycin 5 units/M² IV days 8, 22 Etoposide 60 mg/M² IV days 15. 16 Prednisone 40 mg/M² PO every other day; taper by 10 mg QOD starting at week 10 –decrease vinblastine to 4 mg/M² and vincristine to 1 mg/M² for cycle 3 for patients age > 50 REF: Bartlett et al. J Clin Oncol 1995; 13:1080-1088 PREMEDICATIONS 1. Kytril 1 mg PO/IV 30 minutes before and 12 hours after chemotherapy on days 1 and 15 2. Dexamethasone 20 mg IV 30 minutes before chemotherapy on days 1 and 15 3. Compazine 10 mg PO/IV 30 minutes before chemotherapy on days 8, 16, and 22 OTHER MEDICATIONS Cotrimoxazole DS 1 tablet PO BID until therapy completed Acyclovir 200 mg PO TID until therapy completed Ketoconazole 200 mg PO QD until therapy completed Stool softener daily until therapy completed This is a 12 week regimen (above is repeated every 28 days for 3 cycles)

Continued

Agent	Dosage
	Vincristine—vesicant–avoid extravasation; cumulative neurotoxicity—may produce severe constipation; maximum 2 mg per administration Nitrogen mustard—potent vesicant—avoid extravasation; decomposes rapidly after mixing; must not be mixed in same syringe with any other drug Doxorubicin—monitor cumulative dose for cardiac toxicity (not to exceed 550 mg/M^2 or 450 mg/M^2 with prior chest radiotherapy); vesicant—avoid extravasation; use 50% for bilirubin 1.5-3.0; use 25% for bilirubin > 3.0 Vinblastine—use 50% of dose for bilirubin > 3.0; vesicant—avoid extravasation; watch for neurotoxicity Bleomycin—give test dose of 1-2 units because of possible acute pulmonary, anaphylactoid, or severe febrile reactions; must dose adjust for renal insufficiency; total lifetime dose should not exceed 400 units; avoid high FiO_2 as it can exacerbate pulmonary toxicity
Gemcitabine	Gemcitabine 1250 mg/M^2 IV (over 30 min) days 1,8,15 –20% dose increase permitted if no toxicity after first 4 week cycle REF: Santoro et al. J Clin Oncol 2000; 18:2615-2619 PREMEDICATIONS 1. Compazine 10 mg PO/IV 30 minutes before Repeat every 28 days
Vinblastine	Vinblastine 4-6 mg/M^2 IV day 1 REF: Little et al. J Clin Oncol 1998; 16:584-588 PREMEDICATIONS 1. Compazine 10 mg PO/IV 30 minutes before Repeat every 7-14 days Vinblastine—use 50% of dose for bilirubin > 3.0; vesicant—avoid extravasation; watch for neurotoxicity

Multiple Myeloma

Agent	Dosage			
MP melphalan/ prednisone	Melphalan	10 mg/M²	PO	days 1-4
	Prednisone	60 mg/M²	PO	days 1-4
	REF: Arch Intern Med 1975; 135:147-152			
	OR			
	Melphalan	0.15 mg/kg	PO	days 1-7
	Prednisone	60 mg	PO	days 1-7
	REF: Kyle et al. CRC Crit Rev Oncol/Hematol 1988; 8:93-152			
	–there are numerous variations of the MP regimen			
	Repeat every 28–42 days			
M2 (VBMCP) vincristine/ carmustine (BCNU)/ cyclophos-phamide/ melphalan/ prednisone	Vincristine	0.03 mg/kg	IV	day 1
	BCNU	0.50 mg/kg	IV	day 1
	Cyclophosphamide	10 mg/kg	IV	day 1
	Melphalan	0.25 mg/kg	PO	days 1-4
	Prednisone	1 mg/kg	PO	days 1-7
	then	0.50 mg/kg	PO	days 8-14
	REF: Case et al. Am J Med 1977; 63:897-903			
	PREMEDICATIONS 1. Kytril 1 mg PO/IV 30 minutes before and 12 hours after chemotherapy on day 1			
	OTHER MEDICATIONS 1. Give non-cisplatin delayed-emesis prophylaxis			
	Repeat every 35 days			
	Carmustine—maximum total dose is 1440 mg/M²; causes delayed myelosuppression			
	Vincristine—vesicant–avoid extravasation; cumulative neurotoxicity—may produce severe constipation; maximum 2 mg per administration			
VAD vincristine/ doxorubicin/ dexamethasone	Vincristine	0.4 mg/d	CIV	days 1-4
	Doxorubicin	9 mg/M²/d	CIV	days 1-4
	Dexamethasone	40 mg	PO	days 1-4, 9-12, 17-20
	REF: Barlogie et al. NEJM 1984; 310:1353-1356			

Continued

Brain Cancer | Breast Cancer | Carcinoma of Unknown Primary | Endocrine Cancer | Gastrointestinal Cancer | Genitourinary Cancer | Gynecologic Cancer | Head and Neck Cancer | Hematologic Malignancies

Agent	Dosage
	PREMEDICATIONS 1. Kytril 1 mg POIV 30 minutes before and Q12H during chemotherapy on days 1-4 Repeat every 28 days Doxorubicin—monitor cumulative dose for cardiac toxicity (not to exceed 550 mg/M²); vesicant—avoid extravasation; use 50% for bilirubin 1.5-3.0; use 25% for bilirubin > 3.0
Dexamethasone	Dexamethasone 40 mg PO days 1-4,9-12, 17-20 REF: Alexanian: Ann Intern Med 1986; 105:8-11 Repeat every 35 days
Pamidronate (Aredia)	Pamidronate 90 mg IV day 1 REF: Berenson et al. J Clin Oncol 1998; 16:593-602 Repeat every 28 days
Thalidomide	Thalidomide 200 mg PO QHS daily –dose advanced 200 mg every 2 weeks as tolerated REF: Desikan et al. Blood 1999; 94(Suppl 1):abstract 2685 Thalidomide—providers and pharmacies must be registered with the S.T.E.P.S program; can cause significant somnolence
Waldenstrom's Macroglobulinemia	
	Initial therapy frequently consists of an alkylating agent in conjunction with corticosteroids; these regimens can be found in the CLL (chlorambucil, cyclophosphamide) and multiple myeloma (melphalan) sections.
Cladribine (2-CdA)	2-CdA 0.1 mg/kg/d CIV days 1-7 REF: Dimopoulos et al. J Clin Oncol 1994; 12:2694-2698 Repeat every 28 days for 2 cycles
Fludarabine	Fludarabine 25 mg/M² IV days 1-5 REF: Foran et al. J Clin Oncol 1999; 17:546-553 Repeat every 28 days to maximal response plus 2 cycles Consider prophylaxis with trimethoprim-sulfanethoxazole

Myelodysplastic Syndrome

Agent	Dosage
Cytarabine (ara-c)/ topotecan	Ara-C 1000 mg/M² IV (over 2 h) days 1-5 Topotecan 1.25 mg/M²/d CIV days 1-5 REF: Beran et al. J Clin Oncol 1999; 17:2819-2830 PREMEDICATIONS 1. Kytril 1 mg IV/PO 30 minutes before and 12 hours after chemotherapy 2. Dexamethasone 10 mg IV 30 minutes before chemotherapy OTHER MEDICATIONS –these are given during the period of neutropenia 1. Trimethoprim-sulfamethoxazole DS 1 tab PO BID 2. Fluconazole 100-200 mg PO QD 3. Valacyclovir 500 mg PO QD or Acyclovir 200 mg PO BID Ara-C—high doses can cause CNS toxicity (cerebellar dysfunction); neurotoxicity increases as infusion time increases
7+3 cytarabine/ daunorubicin	See regimen listed under AML
Etoposide (VP-16)—oral	–this regimen has been utilized for CMML VP-16 50 mg PO days 1-21 REF: Doll et al. Leuk Res 1998; 22:7-12 Repeat every 28 days
Thalidomide	Thalidomide 100 mg PO QHS daily REF: Raza et al. Blood 1999; 94(Suppl 1):abstract 2935 Thalidomide—providers and pharmacies must be registered with the S.T.E.P.S program; can cause significant somnolence
Topotecan	Topotecan 2 mg/M²/d CIV days 1-5 REF: Beran et al. Semin Hematol 1998; 35:26-31 PREMEDICATIONS 1. Kytril 1 mg IV/PO 30 minutes before and 12 hours after chemotherapy 2. Dexamethasone 10 mg IV 30 minutes before chemotherapy Repeat every 4-6 weeks for 2 cycles, then adjust to maximum tolerated dose (1-2 mg/M²/d CIV X 5 days) every 4-8 weeks to a maximum of 12 cycles

Non-Hodgkin's Lymphoma

Agent	Dosage
CHOP **cyclophos-phamide/ doxorubicin/ vincristine/ prednisone**	Cyclophosphamide 750 mg/M^2 IV day 1 Doxorubicin 50 mg/M^2 IV day 1 Vincristine 1.4 mg/M^2 IV day 1 Prednisone 100 mg PO days 1-5 REF: McKelvey et al. Cancer 1976; 38:1484-1493 PREMEDICATIONS 1. Kytril 1 mg PO/IV 30 minutes before and 12 hours after chemotherapy 2. Dexamethasone 20 mg IV 30 minutes before chemotherapy Repeat every 21 days Doxorubicin—monitor cumulative dose for cardiac toxicity (not to exceed 550 mg/M^2 or 450 mg/M^2 with prior chest radiotherapy); vesicant—avoid extravasation; use 50% for bilirubin 1.5-3.0; use 25% for bilirubin > 3.0 Vincristine—vesicant–avoid extravasation; cumulative neurotoxicity—may produce severe constipation; maximum 2 mg per administration
CHOP/Rituxin cyclophos-phamide/ doxorubicin/ vincristine/ prednisone/ rituximab	Rituximab 375 mg/M^2 IV day 1 - infusion is started at 50 mg/hr (25 mg/hr in patients with circulating tumor cells) and slowly increased to a maximum of 400 mg/hr (300 mg/hr during initial infusion) Cyclophosphamide 750 mg/M^2 IV day 3 Doxorubican 50 mg/M^2 IV day 3 Vincristine 1.4 mg/M^2 IV day 3 Prednisone 100 mg PO days 3-7 REF: Vose et al. J. Clin Oncol 2001; 19:389-397. Repeat every 21 days OR Rituximab 375 mg/M^2 IV day 1 - infusion is started at 50 mg/hr (25 mg/hr in patients with circulating tumor cells) and slowly increased to a maximum of 400 mg/hr (300 mg/hr during initial infusion) Cyclophosphamide 750 mg/M^2 IV day 1 Doxorubican 50 mg/M^2 IV day 1 Vincristine 1.4 mg/M^2 IV day 1 Prednisone 40 mg/M^2 PO days 1-5

Continued

Agent	Dosage
	OTHER MEDICATIONS: 1. G-CSF 5 mcg/kg SQ days 5-12 REF: Coiffier et al. Blood 2001; 96 (Suppl):abstract 950 Repeat every 21 days for 8 cycles PREMEDICATIONS 1. Kytril 1 mg PO/IV 30 minutes before and 12 hours after chemotherapy 2. Dexamethasone 20 mg IV 30 minutes before chemotherapy 3. Tylenol 650 mg PO 30 minutes before rituximab 4. Benadryl 25 mg PO/IV 30 minutes before rituximab Doxorubicin—monitor cumulative dose for cardiac toxicity (not to exceed 550 mg/M² or 450 mg/M² with prior chest radiotherapy); vesicant—avoid extravasation; use 50% for bilirubin 1.5-3.0; use 25% for bilirubin >3.0 Vincristine—vesicant—avoid extravasation; cumulative neurotoxicity—may produce severe constipation; maximum 2 mg per administration
CVP (COP) cyclophosphamide/ vincristine/ prednisone	Cyclophosphamide 400 mg/M² PO days 1-5 Vincristine 1.4 mg/M² IV day 1 Prednisone 100 mg/M² PO days 1-5 REF: Bagley et al. Ann Intern Med 1972; 76:227-234 —there are many variations of this regimen PREMEDICATIONS 1. Kytril 1 mg PO/IV 30 minutes before and 12 hours after chemotherapy days 1-5 Repeat every 21-28 days Vincristine—vesicant—avoid extravasation; cumulative neurotoxicity—may produce severe constipation; maximum 2 mg per administration
DHAP dexamethasone/ cytarabine/ cisplatin	Cisplatin 100 mg/M² CIV X 24 hr day 1 Cytarabine 2000 mg/M² IV Q12H X 2 doses, each over 3 hr day 2 —start at completion of cisplatin infusion Dexamethasone 40 mg PO/IV days 1-4 REF: Velasquez et al. Blood 1988; 71:117-122

Continued

Agent	Dosage
	PREMEDICATIONS 1. Kytril 1 mg PO/IV 30 minutes before and 12 hours after chemotherapy on days 1 and 2 OTHER MEDICATIONS 1. Give cisplatin delayed-emesis prophylaxis Repeat every 21-28 days Cisplatin—vigorous hydration is required; can be nephrotoxic and ototoxic; can cause peripheral neuropathy; hold or reduce for creatinine > 1.5 Cytarabine—high doses can cause CNS toxicity (cerebellar dysfunction); neurotoxicity increases as infusion time increases
ESHAP **etoposide (VP-16)/ methylpredni-solone/ cytarabine/ cisplatin**	VP-16 — 40 mg/M² — IV (over 1 h) — days 1-4 Methylprednisolone — 500 mg — IV (over 15 min) — days 1-4 Cytarabine — 2000 mg/M² — IV (over 2 h) — day 5 Cisplatin — 25 mg/M² — CIV (over 96 h) — days 1-4 REF: Velasquez et al. J Clin Oncol 1994; 12:1169-1176 PREMEDICATIONS 1. Kytril 1 mg PO/IV Q12H for 10 doses, starting 30 minutes before chemotherapy on day 1 OTHER MEDICATIONS 1. Give cisplatin delayed-emesis prophylaxis Repeat every 21-28 days Cisplatin—vigorous hydration is required; can be nephrotoxic and ototoxic; can cause peripheral neuropathy; hold or reduce for creatinine > 1.5 Cytarabine—high doses can cause CNS toxicity (cerebellar dysfunction); neurotoxicity increases as infusion time increases
ICE **ifosfamide/ carboplatin/ etoposide (VP-16)**	–also used as a stem cell mobilization regimen (with G-CSF at 10 mcg/kg/d) Ifosfamide — 5000 mg/M² — CIV X 24 hr — day 2 Mesna — 5000 mg/M² — CIV X 24 hr — day 2 Carboplatin — AUC 5 — IV — day 2 VP-16 — 100 mg/M² — IV — days 1-3 REF: Moskowitz et al. J Clin Oncol 1999; 17:3776-3785 PREMEDICATIONS 1. Kytril 1 mg PO/IV 30 minutes before and 12 hours after chemotherapy on days 1-3 2. Dexamethasone 20 mg IV 30 minutes before chemo-therapy on day 2

Continued

Agent	Dosage			
	OTHER MEDICATIONS 1. G-CSF 5 mcg/kg/d SQ days 5-12 2. Give non-cisplatin delayed emesis prophylaxis Repeat every 21 days Ifosfamide—adequate hydration is necessary to prevent nephrotoxicity			
MINE mesna/ ifosfamide/ mitoxantrone/ etoposide (VP-16)	Mesna	1333 mg/M^2	IV at same time as ifosfamide	days 1-3
	Mesna	500 mg	PO 4 hr after ifosfamide	days 1-3
	Ifosfamide	1333 mg/M^2	IV (over 1 h)	days 1-3
	Mitoxantrone	8 mg/M^2	IV (over 15 min)	day 1
	VP-16	65 mg/M^2	IV (over 1 h)	days 1-3
	REF: Rodriguez et al. J Clin Oncol 1995; 13:1734-1741 PREMEDICATIONS 1. Kytril 1 mg PO/IV 30 minutes before and 12 hours after chemotherapy on days 1-3 2. Dexamethasone 20 mg IV 30 minutes before chemotherapy on days 1-3 Repeat every 21-28 days Mitoxantrone—watch cumulative dose—do not exceed 140 mg/M^2; possible cardiac toxicity Ifosfamide—adequate hydration is necessary to prevent nephrotoxicity			
MINE/ESHAP	–MINE regimen as above, to a maximum of 6 cycles; this is followed by ESHAP as above (with exception of increase of VP-16-60 mg/M^2/d for 4 days) for 3 cycles if there was a complete response to MINE and 6 cycles if there was a partial response (or no response) to MINE –antiemetics and warnings are as listed with the individual regimens REF: Rodriguez et al. J Clin Oncol 1995; 13:1734-1741			

Brain Cancer | Breast Cancer | Carcinoma of Unknown Primary | Endocrine Cancer | Gastrointestinal Cancer | Genitourinary Cancer | Gynecologic Cancer | Head and Neck Cancer | Hematologic Malignancies

Agent	Dosage			
MACOP-B methotrexate/ doxorubicin/ cyclophos-phamide/ vincristine/ bleomycin/ prednisone/ folinic acid	Methotrexate	400 mg/M²	IV	days 8,36,64 (weeks 2,6,10)
	–100 mg/M² bolus in 20 minutes, then 300 mg/M² as 2 hr infusion			
	Folinic Acid	15 mg	PO Q6H X 6 doses starting 24 hours after methotrexate	
	Doxorubicin	50 mg/M²	IV	days 1,15,29,43, 57,71 (weeks 1,3,5,7,9,11)
	Cyclophosphamide	350 mg/M²	IV	days 1,15,29,43, 57,71 (weeks 1,3,5,7,9,11)
	Vincristine	1.4 mg/M²	IV	days 8,22,36,50, 64,78 (weeks 2,4,6,8,10,12)
	Bleomycin	10 mg/M²	IV	days 22,50,78 (weeks 4,8,12)
	Prednisone	75 mg	PO	daily for 12 weeks (tapered over last 14 days)

REF: Schneider et al. J Clin Oncol 1990; 8:94-102

PREMEDICATIONS
1. Hydrocortisone 100 mg IV given prior to each dose of Bleomycin
2. Kytril 1 mg PO/IV 30 minutes before and 12 hours after chemotherapy on days 1, 8, 15, 29, 36, 43, 57, 64 and 71
3. Dexamethasone 20 mg IV 30 minutes before chemotherapy on days 1, 8, 15, 29, 36, 43, 57, 64 and 71

OTHER MEDICATIONS
1. Trimethoprim-sulfamethoxazole 2 DS tablet PO BID daily for 12 weeks
2. Ketoconazole 200 mg PO daily for 12 weeks

Cycle is given only one time, over a 12 week period

Doxorubicin—monitor cumulative dose for cardiac toxicity (not to exceed 550 mg/M² or 450 mg/M² with prior chest radiotherapy); vesicant—avoid extravasation; use 50% for bilirubin 1.5-3.0; use 25% for bilirubin > 3.0

	Agent	Dosage		

Vincristine—vesicant–avoid extravasation; cumulative neurotoxicity—may produce severe constipation; maximum 2 mg per administration

Bleomycin—give test dose of 1-2 units because of possible acute pulmonary, anaphylactoid, or severe febrile reactions; must dose adjust for renal insufficiency; total lifetime dose should not exceed 400 units; avoid high FiO_2 as it can exacerbate pulmonary toxicity

Methotrexate—use 75% dose for CrCl < 50; 50% dose if CrCl < 25; do not give if patient has an effusion ("reservoir effect")

Regimen	Agent	Dosage	Route	Schedule
m-BACOD bleomycin/ doxorubicin/ cyclophosphamide/ vincristine/ dexamethasone/ methotrexate/ folinic acid	Bleomycin	4 mg/M²	IV	day 1
	Doxorubicin	45 mg/M²	IV	day 1
	Cyclophosphamide	600 mg/M²	IV	day 1
	Vincristine	1 mg/M²	IV	day 1
	Dexamethasone	6 mg/M²	PO	days 1-5
	Methotrexate	200 mg/M²	IV (over 1 h)	days 8,15
	Folinic Acid	10 mg/M²	PO Q6H X 8 doses	
	–starting 24 hours after methotrexate			

REF: Shipp et al. J Clin Oncol 1990; 8:84-93

PREMEDICATIONS
1. Kytril 1 mg PO/IV 30 minutes before and 12 hours after chemotherapy on days 1, 8, and 15
2. Dexamethasone 20 mg IV 30 minutes before chemotherapy on days 1, 8, and 15

Repeat every 21 days for up to 10 cycles

Doxorubicin—monitor cumulative dose for cardiac toxicity (not to exceed 550 mg/M² or 450 mg/M² with prior chest radiotherapy); vesicant—avoid extravasation; use 50% for bilirubin 1.5-3.0; use 25% for bilirubin > 3.0

Vincristine—vesicant–avoid extravasation; cumulative neurotoxicity—may produce severe constipation; maximum 2 mg per administration

Bleomycin—give test dose of 1-2 units because of possible acute pulmonary, anaphylactoid, or severe febrile reactions; must dose adjust for renal insufficiency; total lifetime dose should not exceed 400 units; avoid high FiO_2 as it can exacerbate pulmonary toxicity

Methotrexate—use 75% dose for CrCl < 50; 50% dose if CrCl < 25; do not give if patient has an effusion ("reservoir effect")

Agent	Dosage			
Low-dose m-BACOD bleomycin/ doxorubicin/ cyclophos-phamide/ vincristine/ dexamethasone/ methotrexate/ folinic acid/ cytarabine (ara-C)	–this regimen is utilized in AIDS associated lymphomas			
	Bleomycin	4 mg/M²	IV	day 1
	Doxorubicin	25 mg/M²	IV	day 1
	Cyclophosphamide	300 mg/M²	IV	day 1
	Vincristine	1.4 mg/M²	IV	day 1
	Dexamethasone	3 mg/M²	PO	days 1-5
	Methotrexate	200 mg/M²	IV (over 1 h)	day 15
	Folinic Acid	10 mg/M²	PO Q6H X 8 doses starting 24 hr after methotrexate	
	Ara-C	50 mg	IT (intrathecal)	days 1,8,15,22

REF: Kaplan et al. NEJM 1997; 336:1641-1648

PREMEDICATIONS

1. Kytril 1 mg PO/IV 30 minutes before and 12 hours after chemotherapy on days 1, 8, and 15
2. Dexamethasone 20 mg IV 30 minutes before chemotherapy on days 1, 8, and 15

Repeat every 21 days

Doxorubicin—monitor cumulative dose for cardiac toxicity (not to exceed 550 mg/M² or 450 mg/M² with prior chest radiotherapy); vesicant—avoid extravasation; use 50% for bilirubin 1.5-3.0; use 25% for bilirubin > 3.0

Vincristine—vesicant–avoid extravasation; cumulative neurotoxicity—may produce severe constipation; maximum 2 mg per administration

Bleomycin—give test dose of 1-2 units because of possible acute pulmonary, anaphylactoid, or severe febrile reactions; must dose adjust for renal insufficiency; total lifetime dose should not exceed 400 units; avoid high FiO_2 as it can exacerbate pulmonary toxicity

Methotrexate—use 75% dose for CrCl < 50; 50% dose if CrCl < 25; do not give if patient has an effusion ("reservoir effect")

Agent	Dosage			
ProMACE-CytaBOM prednisone/ doxorubicin/ cyclophos-phamide/ etoposide/ cytarabine/ bleomycin/ vincristine/ methotrexate/ folinic acid	Prednisone	60 mg/M²	PO	days 1-14
	Doxorubicin	25 mg/M²	IV	day 1
	Cyclophosphamide	650 mg/M²	IV	day 1
	Etoposide	120 mg/M²	IV	day 1
	Cytarabine	300 mg/M²	IV	day 8
	Bleomycin	5 mg/M²	IV	day 8
	Vincristine	1.4 mg/M²	IV	day 8
	Methotrexate	120 mg/M²	IV	day 8
	Folinic acid	25 mg/M²	PO	Q6H for 4 doses starting 24h after methotrexate
	REF: Longo et al. J Clin Oncol 1991; 9:25-38 PREMEDICATIONS 1. Kytril 1 mg PO/IV 30 minutes before and 12 hours after chemotherapy on days 1 and 8 2. Dexamethasone 20 mg IV before chemotherapy on days 1 and 8 OTHER MEDICATIONS 1. Trimethoprim-sulfamethoxazole DS one tablet BID Repeat every 21 days for at least 6 cycles (2 cycles beyond CR) Doxorubicin—monitor cumulative dose for cardiac toxicity (not to exceed 550 mg/M² or 450 mg/M² with prior chest radiotherapy); vesicant—avoid extravasation; use 50% for bilirubin 1.5-3.0; use 25% for bilirubin > 3.0 Vincristine—vesicant–avoid extravasation; cumulative neurotoxicity—may produce severe constipation; maximum 2 mg per administration Bleomycin—give test dose of 1-2 units because of possible acute pulmonary, anaphylactoid, or severe febrile reactions; must dose adjust for renal insufficiency; total lifetime dose should not exceed 400 units; avoid high FiO_2 as it can exacerbate pulmonary toxicity Methotrexate—use 75% dose for CrCl < 50; 50% dose if CrCl < 25; do not give if patient has an effusion ("reservoir effect")			
Gemcitabine	Gemcitabine	1250 mg/M²	IV	days 1,8,15
	REF: Fossa et al. J Clin Oncol 1999; 17:3786-3792 PREMEDICATIONS 1. Compazine 10 mg PO/IV 30 minutes before chemotherapy Repeat every 28 days			

PRIMARY CNS LYMPHOMAS

Agent	Dosage
Methotrexate/ Radiotherapy	Methotrexate 1 gm/M² IV (over 6 h) days 1,8 Leucovorin 15 mgPO Q6H for 72 hr –start 24 hours after start of Methotrexate Ara-C 60 mg IT BIW for 3 wks –then weekly for 3 doses after clearance of CSF REF: O'Brien, et al. J Clin Oncol 2000; 18: 519-526 PREMEDICATIONS 1. Kytril 1 mg PO/IV 30 minutes before and 12 hours after chemotherapy on days 1 and 8 2. Dexamethasone 20 mg IV before chemotherapy on days 1 and 8 Radiotherapy—4500 cGy in 25 fractions, followed by 5.4 Gy to isocenter; starts on day 15 –spinal Radiotherapy to 36 Gy in 24 fractions if cytology is positive

CUTANEOUS T-CELL LYMPHOMAS

Agent	Dosage
Bexarotene (Targretin)	–for use in cutaneous T-cell lymphomas Bexarotene 300 mg/M²/d PO daily REF: Duvic et al. Blood 1999; 94(Suppl 1):abstract 2927 Bexarotene—causes severe hyperlipidemia in majority of patients treated; may require concomitant lipid-lowering therapy
Denileukin diftitox (Ontak)	–for use in refractory CD25 positive cutaneous T-cell lymphomas Ontak 9-18 µg/kg IV(over 15 min) days 1-5 REF: PDR/package insert PREMEDICATIONS 1. Diphenhydramine 25-50 mg PO/IV 30 minutes before treatment 2. Tylenol 650 mg PO 30 minutes before treatment Repeat every 21 days Ontak—watch for high incidence of acute hypersensitivity reactions; be prepared to treat possible anaphylaxis
Gemcitabine	Gemcitabine 1200 mg/M² IV (over 30 min) days 1,8,15 REF: Zinzani et al. J Clin Oncol 2000; 18:2603-2606 PREMEDICATIONS 1. Compazine 10 mg PO/IV 30 minutes before Repeat every 28 days

LOW-GRADE NON-HODGKIN'S LYMPHOMAS

Agent	Dosage
	Please refer to regimens outlined in the CLL section
Cladribine (2-CdA) mitoxantrone	–as therapy for low-grade or mantle cell lymphoma 2-CdA 5 mg/M^2 IV days 1-3 Mitoxantrone 8 mg/M^2 IV days 1-2 –mitoxantrone dose is reduced to 12 mg/M^2 on day 1 only if previously treated REF: Rummel et al. Blood 1999; 94(Suppl 1):abstract 2931 PREMEDICATIONS 1. Compazine 10 mg PO/IV 30 minutes before chemotherapy on days 1-3 Repeat every 28 days
FND fludarabine/ mitoxantrone/ dexamethasone	Fludarabine 25 mg/M^2 IV days 1-3 Mitoxantrone 10 mg/M^2 IV day 1 Dexamethasone 20 mg PO/IV days 1-5 REF: McLaughlin et al. J Clin Oncol 1996: 14:1262-1268 PREMEDICATIONS 1. Kytril 1 mg PO/IV 30 minutes before chemotherapy on day 1 2. Compazine 10 mg PO/IV before chemotherapy on days 2 and 3 OTHER MEDICATIONS 1. Trimethoprim-sulfamethoxazole DS 1 tablet BID for prophylaxis Repeat every 28 days Mitoxantrone—watch cumulative dose—do not exceed 140 mg/M^2; possible cardiac toxicity
Cladribine (2-CdA)	–as therapy for mantle cell lymphoma 2-CdA 5 mg/M^2 IV days 1-5 REF: Inwards et al. Blood 1999; 94(Suppl 1):abstract 2930 PREMEDICATIONS 1. Compazine 10 mg PO/IV 30 minutes before chemotherapy on days 1-5 Repeat every 28 days for 2-6 cycles

Agent	Dosage
Fludarabine	Fludarabine 25 mg/M² IV days 1-5 REF: Redman et al. J Clin Oncol 1992; 10:790-794 PREMEDICATIONS 1. Compazine 10 mg PO/IV 30 minutes before chemotherapy on days 1-5 Repeat every 21-28 days Consider prophylaxis with trimethoprim-sulfanethoxazole
Rituximab	Rituximab 375 mg/M² IV days 1,8,15,22 –infusion is started at 50 mg/hr (25 mg/hr in patients with circulating tumor cells) and slowly increased to a maximum of 400 mg/hr (300 mg/hr during initial infusion) REF: McLaughlin et al. J Clin Oncol 1998; 16:2825-2833 PREMEDICATIONS 1. Tylenol 650 mg PO 30 minutes before 2. Benadryl 25 mg PO/IV 30 minutes before

Chapter 10
Lung Cancer

- Mesothelioma
- Non-Small-Cell Lung Cancer
- Small-Cell Lung Cancer

Mesothelioma

Cisplatin/ gemcitabine	Cisplatin	100 mg/M²	IV	day 1
	Gemcitabine	1000 mg/M²	IV	days 1,8,15

REF: Byrne et al. J Clin Oncol 1999; 17:25-30

PREMEDICATIONS
1. Kytril 1 mg PO/IV 30 minutes before and 12 hours after chemotherapy on day 1
2. Dexamethasone 20 mg IV 30 minutes before chemotherapy on day 1
3. Compazine 10 mg PO/IV 30 minutes before chemotherapy on days 8 and 15

OTHER MEDICATIONS
1. Give cisplatin delayed–emesis prophylaxis

Repeat every 28 days

Cisplatin—vigorous hydration is required; can be nephrotoxic and ototoxic; can cause peripheral neuropathy; hold or reduce for creatinine > 1.5

Cisplatin/ mitomycin C	Cisplatin	75 mg/M²	IV	day 1
	Mitomycin C	10 mg/M²	IV	day 1

REF: Chahinian et al. J Clin Oncol 1993; 11:1559-1565

PREMEDICATIONS
1. Kytril 1 mg PO/IV 30 minutes before and 12 hours after chemotherapy
2. Dexamethasone 20 mg IV 30 minutes before chemotherapy

OTHER MEDICATIONS
1. Give cisplatin delayed–emesis prophylaxis

Repeat every 28 days

Cisplatin—vigorous hydration is required; can be nephrotoxic and ototoxic; can cause peripheral neuropathy; hold or reduce for creatinine > 1.5

Mitomycin C—myelosuppression occurs late (approximately 4 weeks); limit cumulative dose to 50 mg/M² (vascular toxicity)

Agent	Dosage			
Cyclophos-phamide/ doxorubicin/ cisplatin	Cyclophosphamide	500 mg/M^2	IV	day 1
	Doxorubicin	50 mg/M^2	IV	day 1
	Cisplatin	80 mg/M^2	IV	day 1
	–cisplatin dose reduced to 50 mg/M^2 after 1st cycle			

REF: Shin et al. Cancer 1995; 76:2230-2236

PREMEDICATIONS

1. Kytril 1 mg PO/IV 30 minutes before and 12 hours after chemotherapy
2. Dexamethasone 20 mg IV 30 minutes before chemotherapy

OTHER MEDICATIONS

1. Give cisplatin delayed-emesis prophylaxis

Repeat every 21 days

Cisplatin—vigorous hydration is required; can be nephrotoxic and ototoxic; can cause peripheral neuropathy; hold or reduce for creatinine > 1.5

Doxorubicin—monitor cumulative dose for cardiac toxicity (not to exceed 550 mg/M^2 or 450 mg/M^2 with prior chest radiotherapy); vesicant—avoid extravasation; use 50% for bilirubin 1.5-3.0; use 25% for bilirubin > 3.0

Non-Small-Cell Lung Cancer

CP carboplatin/ paclitaxel	Paclitaxel	225 mg/M²	IV (over 3 h)	day 1
	–followed by			
	Carboplatin	AUC 6	IV (over 1 h)	day 1

REF: Kelly et al. Proc Amer Soc Clin Onc 1999; abstract 1777

OR

Paclitaxel	175 mg/M²	IV (over 3 h)	day 1
–followed by			
Carboplatin	AUC 7	IV (over 1 h)	day 1

REF: Kosmidis et al. Ann Oncol 1997; 8:697-699

PREMEDICATIONS
1. Dexamethasone 20 mg IV 30 minutes before chemotherapy
   OR
   Dexamethasone 20 mg PO 6 and 12 hours prior
2. Diphenhydramine 50 mg IV 30 minutes before chemotherapy
3. Cimetidine 300 mg IV 30 minutes before chemotherapy
4. Kytril 1 mg PO/IV 30 minutes before and 12 hours after chemotherapy

OTHER MEDICATIONS
1. Give cisplatin delayed–emesis prophylaxis
2. Dexamethasone 4 mg PO BID for 6 doses after chemotherapy (for myalgias)

Repeat every 21 days

Cisplatin/ vinblastine	–followed by XRT			
	Vinblastine	5 mg/M²	IV	days 1,8,15,22,29
	Cisplatin	100 mg/M²	IV	days 1,29

–radiotherapy is started on day 50, to 60 Gy over a 6 week period

REF: Dillman et al. NEJM 1990; 323:940-945

PREMEDICATIONS
1. Kytril 1 mg PO/IV 30 minutes before and 12 hours after chemotherapy on days 1 and 29
2. Dexamethasone 20 mg IV 30 minutes before chemotherapy on days 1 and 29
3. Compazine 10 mg PO/IV 30 minutes before chemotherapy on days 8, 15, and 22

Continued

OTHER MEDICATIONS
1. Give cisplatin delayed–emesis prophylaxis

Cisplatin—vigorous hydration is required; can be nephrotoxic and ototoxic; can cause peripheral neuropathy; hold or reduce for creatinine > 1.5

Vinblastine—use 50% of dose for bilirubin > 3.0; vesicant–avoid extravasation

Docetaxel/ cisplatin

Docetaxel	75 mg/M²	IV	day 1
cisplatin	75 mg/M²	IV	day 1

REF: Schiller et al. Proc ASCO 2000:abstract 2

PREMEDICATIONS
1. Kytril 1 mg PO/IV 30 minutes before and 12 hours after chemotherapy on day 1
2. Dexamethasone 20 mg IV 30 minutes before chemotherapy on day 1
3. Cemitidine 300 mg IV 30 minutes before chemotherapy
4. Diphenhydramine 25-50 mg IV 30 minutes before chemotherapy

OTHER MEDICATIONS
1. Give cisplatin delayed-emesis prophylaxis
2. Dexamethasone 8 mg PO BID for 8 doses—start day prior to chemo (decreases lower extremity edema)

Repeat every 21 days

Cisplatin—vigorous hydration is required; can be nephrotoxic and ototoxic; can cause peripheral neuropathy; hold or reduce for creatinine > 1.5

EP (PE) cisplatin/ etoposide

Etoposide	100 mg/M²	IV	days 1-3
Cisplatin	100 mg/M²	IV	day 1

REF: Cardenal et al. J Clin Oncol 1999; 17:12-18

–there are multiple variants of this regimen

PREMEDICATIONS
1. Kytril 1 mg PO/IV 30 minutes before and 12 hours after chemotherapy on day 1
2. Dexamethasone 20 mg IV 30 minutes before chemotherapy on day 1
3. Compazine 10 mg PO/IV 30 minutes before chemotherapy on days 2 and 3

OTHER MEDICATIONS
1. Give cisplatin delayed-emesis prophylaxis

Repeat every 21-28 days

Continued

Agent	Dosage			
	Cisplatin—vigorous hydration is required; can be nephrotoxic and ototoxic; can cause peripheral neuropathy; hold or reduce for creatinine > 1.5			
Gemcitabine/ cisplatin	Gemcitabine	1000 mg/M²	IV	days 1,8,15
	Cisplatin	100 mg/M²	IV	day 1
	REF: Sandler et al. J Clin Oncol 2000; 18:122-130 PREMEDICATIONS 1. Kytril 1 mg PO/IV 30 minutes before and 12 hours after chemotherapy on day 1 2. Dexamethasone 20 mg IV 30 minutes before chemotherapy on day 1 3. Compazine 10 mg PO/IV 30 minutes before chemotherapy on days 8 and 15 OTHER MEDICATIONS 1. Give cisplatin delayed–emesis prophylaxis Repeat every 28 days Cisplatin—vigorous hydration is required; can be nephrotoxic and ototoxic; can cause peripheral neuropathy; hold or reduce for creatinine > 1.5 Gemcitabine—dosage modifications are based on degree of thrombocytopenia or neutropenia			
Gemcitabine/ vinorelbine	Gemcitabine	1200 mg/M²	IV	days 1, 8
	Vinorelbine	30 mg/M²	IV	days 1, 8
	REF: Lorusso et al. J Clin Oncol 2000; 405-411 PREMEDICATIONS 1. Kytril 1 mg PO/IV 30 minutes before and 12 hours after chemotherapy Repeat every 21 days Vinorelbine—vesicant; avoid extravasation; can cause peripheral neuropathy			
MVP mitomycin C/ vinblastine/ cisplatin	Mitomycin C	8 mg/M²	IV	day 1 (of every other course)
	Vinblastine –maximum dose is 10 mg	6 mg/M²	IV	day 1
	Cisplatin	50 mg/M²	IV	day 1
	REF: Ellis et al. Br J Cancer 1995; 71:366-370			

Continued

Agent	Dosage
	PREMEDICATIONS 1. Kytril 1 mg PO/IV 30 minutes before and 12 hours after chemotherapy 2. Dexamethasone 20 mg IV 30 minutes before chemotherapy OTHER MEDICATIONS 1. Give cisplatin delayed-emesis prophylaxis Repeat every 21 days Cisplatin—vigorous hydration is required; can be nephrotoxic and ototoxic; can cause peripheral neuropathy; hold or reduce for creatinine > 1.5 Vinblastine—use 50% of dose for bilirubin > 3.0; vesicant–avoid extravasation Mitomycin C—myelosuppression occurs late (approximately 4 weeks); limit cumulative dose to 50 mg/M² (vascular toxicity)
VC vinorelbine/ cisplatin	Vinorelbine 25 mg/M² IV days 1,8,15,22 Cisplatin 100 mg/M² IV day 1 REF: Kelly et al. Proc Amer Soc Clin Onc 1999; abstract 1777 PREMEDICATIONS 1. Kytril 1 mg PO/IV 30 minutes before and 12 hours after chemotherapy on day 1 2. Dexamethasone 20 mg IV 30 minutes before chemotherapy on day 1 3. Compazine 10 mg PO/IV 30 minutes before chemotherapy on days 8, 15, and 22 OTHER MEDICATIONS 1. Give cisplatin delayed-emesis prophylaxis Repeat every 28 days Vinorelbine—vesicant; avoid extravasation; can cause peripheral neuropathy Cisplatin—vigorous hydration is required; can be nephrotoxic and ototoxic; can cause peripheral neuropathy; hold or reduce for creatinine > 1.5
Docetaxel	Docetaxel 100 mg/M² IV(over 1 h) day 1 REF: Gandara et al. J Clin Oncol 2000; 18:131-135 OR Docetaxel 75 mg/M² IV(over 1 h) day 1 REF: Fossella et al. J Clin Oncol 2000; 18:2354-2362

Agent	Dosage			
	PREMEDICATIONS 1. Dexamethasone 20 mg IV 30 minutes before chemotherapy 2. Cimetidine 300 mg IV 30 minutes before chemotherapy 3. Diphenhydramine 25-50 mg IV 30 minutes before chemotherapy 4. Compazine 10 mg PO/IV 30 minutes before chemotherapy OTHER MEDICATIONS 1. Dexamethasone 8 mg PO BID for 8 doses—start day prior to chemo (decreases lower extremity edema) Repeat every 21 days			
Etoposide (VP-16)–oral	Etoposide	100 mg	PO	days 1-7
	Etoposide	100 mg	PO QOD	days 8-14
	REF: Kakolyris et al. Am J Clin Oncol 1998; 21:505-508 Repeat every 28 days			
Gemcitabine	Gemcitabine	1000 mg/M²	IV	days 1,8,15
	REF: Crino et al. J Clin Oncol 1999; 17:2081-2085 PREMEDICATIONS 1. Compazine 10 mg PO/IV 30 minutes before chemotherapy Repeat every 28 days			
Topotecan	Topotecan	1.5 mg/M²/d	IV (over 30 min)	days 1-5
	REF: Perez-Soler et al. J Clin Oncol 1996; 14:503-13 PREMEDICATIONS 1. Kytril 1 mg PO/IV 30 minutes before and 12 hours after chemotherapy on days 1-5 2. Dexamethasone 20 mg IV 30 minutes before chemotherapy on days 1-5 Repeat every 21 days Topotecan—hold for ANC < 1500 or platelets < 100,000; decrease dose by 0.25 mg/M²/d for prior episode of severe neutropenia or administer G-CSF starting on day 6			
Vinorelbine	Vinorelbine	30 mg/M²	IV	every 7 days
	–decrease dose to 15 mg/M² when ANC 1000-1499 REF: Crawford et al. J Clin Oncol 1996; 14:2774-2784 PREMEDICATIONS 1. Compazine 10 mg PO/IV 30 minutes before chemotherapy Repeat every 7 days Vinorelbine—vesicant; avoid extravasation; can cause peripheral neuropathy			

Small-Cell Lung Cancer

Agent	Dosage
Carboplatin/ paclitaxel	Paclitaxel 175 mg/M² IV (over 3 h) day 1 –followed by Carboplatin AUC 7 IV (over 3 h) day 1 REF: Groen et al. J Clin Oncol 1999; 17:927-932 PREMEDICATIONS 1. Dexamethasone 20 mg IV 30 minutes before chemotherapy OR Dexamethasone 20 mg PO 6 and 12 hours prior 2. Diphenhydramine 50 mg IV 30 minutes before chemotherapy 3. Cimetidine 300 mg IV 30 minutes before chemotherapy 4. Kytril 1 mg PO/IV 30 minutes before and 12 hours after chemotherapy OTHER MEDICATIONS 1. Give non-cisplatin delayed-emesis prophylaxis 2. Dexamethasone 4 mg PO BID for 6 doses after paclitaxel (for myalgias) Repeat every 21 days
CAE (ACE) cyclophosphamide/ doxorubicin/ etoposide	Cyclophosphamide 1000 mg/M² IV day 1 Doxorubicin 45 mg/M² IV day 1 Etoposide (VP-16) 50 mg/M² IV days 1-5 REF: Aisner et al. Semin Oncol 1986; 13:54-62 PREMEDICATIONS 1. Kytril 1 mg PO/IV 30 minutes before and 12 hours after chemotherapy on day 1 2. Dexamethasone 20 mg IV 30 minutes before chemotherapy on day 1 3. Compazine 10 mg PO/IV 30 minutes before chemotherapy on days 2-5 OTHER MEDICATIONS 1. May need to give non-cisplatin delayed-emesis prophylaxis Repeat every 21 days Doxorubicin—monitor cumulative dose for cardiac toxicity (not to exceed 550 mg/M² or 450 mg/M² with prior chest radiotherapy); vesicant—avoid extravasation; use 50% for bilirubin 1.5-3.0; use 25% for bilirubin > 3.

Agent	Dosage
CAV cyclophos-phamide/ doxorubicin/ vincristine	Cyclophosphamide 1000 mg/M² IV day 1 Doxorubicin 40 mg/M² IV day 1 Vincristine 1 mg/M² IV day 1 REF: Roth et al. J Clin Oncol 1992; 10:282-291 PREMEDICATIONS 1. Kytril 1 mg PO/IV 30 minutes before and 12 hours after chemotherapy 2. Dexamethasone 20 mg IV 30 minutes before chemo-therapy OTHER MEDICATIONS 1. May need to give non-cisplatin delayed-emesis prophylaxis Repeat every 21 days Doxorubicin—monitor cumulative dose for cardiac toxicity (not to exceed 550 mg/M² or 450 mg/M² with prior chest radiotherapy); vesicant—avoid extravasation; use 50% for bilirubin 1.5-3.0; use 25% for bilirubin > 3.0 Vincristine—vesicant–avoid extravasation; cumulative neurotoxic-ity—may produce severe constipation; maximum 2 mg per administration
EC etoposide/ carboplatin	Etoposide 120 mg/M² IV days 1-3 Carboplatin AUC 6 IV day 1 REF: Birch et al. Semin Oncol 1997; 24(4 Suppl 12):135-137 PREMEDICATIONS 1. Kytril 1 mg PO/IV 30 minutes before and 12 hours after chemotherapy on day 1 2. Dexamethasone 20 mg IV 30 minutes before chemo-therapy on day 1 3. Compazine 10 mg PO/IV 30 minutes before chemotherapy on days 2 and 3 OTHER MEDICATIONS 1. Give cisplatin delayed-emesis prophylaxis Repeat every 28-35 days
EP (PE) cisplatin/ etoposide	Etoposide 100 mg/M² IV days 1-3 Cisplatin 25 mg/M² IV days 1-3 REF: Loehrer et al. Semin Oncol 1988; 15:2-8 –multiple variants of this regimen have been published

Continued

Agent	Dosage
	PREMEDICATIONS 1. Kytril 1 mg PO/IV 30 minutes before and 12 hours after chemotherapy on days 1-3 2. Dexamethasone 20 mg IV 30 minutes before chemotherapy on days 1-3 OTHER MEDICATIONS 1. Give cisplatin delayed-emesis prophylaxis Repeat every 21 days Cisplatin—vigorous hydration is required; can be nephrotoxic and ototoxic; can cause peripheral neuropathy; hold or reduce for creatinine > 1.5
Irinotecan/ cisplatin	Irinotecan 60 mg/M^2 IV days 1, 8, 15 Cisplatin 60 mg/M^2 IV day 1 REF: Kudoh et al. J Clin Oncol 1998; 1068-1074 PREMEDICATIONS 1. Kytril 1 mg PO/IV 30 minutes before and 12 hours after chemotherapy on days 1, 8, 15 2. Dexamethasone 20 mg IV 30 minutes before chemotherapy on days 1, 8, 15 OTHER MEDICATIONS 1. Give cisplatin delayed-emesis prophylaxis 2. Lomotil 4 mg PO at first sign of any loose stool and 2 mg every 2 hours until formed stool Repeat every 28 days for 4 (with XRT in limited disease) or 6 (extensive disease) cycles Cisplatin—vigorous hydration is required; can be nephrotoxic and ototoxic; can cause peripheral neuropathy; hold or reduce for creatinine > 1.5
PCE cyclophos-phamide/ doxorubicin/ etoposide (VP-16)	Paclitaxel 200 mg/M^2 IV (over 1 h) day 1 Carboplatin AUC 6 IV day 1 VP-16 50 mg PO QOD days 1-10 –alternating with VP-16 100 mg PO QOD days 1-10 –if limited stage, concurrent XRT to 45 Gy is given with cycles 3 and 4 REF: Hainsworth et al. J Clin Oncol 1997; 15:3464-3470 PREMEDICATIONS 1. Dexamethasone 20 mg IV 30 minutes before chemotherapy OR

Continued

Agent	Dosage
	Dexamethasone 20 mg PO 6 and 12 hours prior 2. Diphenhydramine 50 mg IV 30 minutes before chemotherapy 3. Cimetidine 300 mg IV 30 minutes before chemotherapy 4. Kytril 1 mg PO/IV 30 minutes before and 12 hours after chemotherapy OTHER MEDICATIONS 1. Give cisplatin delayed-emesis prophylaxis 2. Dexamethasone 4 mg PO BID for 6 doses after chemotherapy (for myalgias) Repeat every 21 days
PE/XRT cisplatin/ etoposide (VP-16)/ concurrent radiotherapy	Cisplatin 60 mg/M^2 IV day 1 VP-16 120 mg/M^2 IV days 1-3 –radiotherapy to 45 Gy is given, starting concurrently with cycle 1 of chemotherapy –a total of 4 cycles of chemotherapy are given, 2 during radiotherapy and 2 after REF: Turrisi et al. NEJM 1999; 340:265-271 PREMEDICATIONS 1. Kytril 1 mg PO/IV 30 minutes before and 12 hours after cisplatin 2. Dexamethasone 20 mg IV 30 minutes before cisplatin 3. Compazine 10 mg PO/IV 30 minutes before etoposide OTHER MEDICATIONS 1. Give cisplatin delayed-emesis prophylaxis Repeat every 21 days Cisplatin—vigorous hydration is required; can be nephrotoxic and ototoxic; Can cause peripheral neuropathy; hold or reduce for creatinine > 1.5
Etoposide (VP-16)–oral	Etoposide 100 mg PO days 1-21 REF: Sessa et al. Ann Oncol 1993; 4:553-558 Repeat every 28 days
Gemcitabine	Gemcitabine 1000-1250 mg/M^2 IV days 1,8,15 REF: Cormier et al. Ann Oncol 1994; 5:283-285 PREMEDICATIONS 1. Compazine 10 mg PO/IV 30 minutes before chemotherapy Repeat every 28 days

Agent	Dosage
Topotecan	Topotecan 1.5 mg/M²/d IV (over 30 min) days 1-5 REF: Ardizonni et al. J Clin Oncol 1997; 15:2090-2096 PREMEDICATIONS 1. Kytril 1 mg PO/IV 30 minutes before and 12 hours after chemotherapy on day 1 2. Dexamethasone 20 mg IV 30 minutes before chemotherapy on day 1 Repeat every 21 days Topotecan—hold for ANC < 1500 or platelets < 100,000; decrease dose by 0.25 mg/M²/d for prior episode of severe neutropenia or administer G-CSF starting on day 6

Chapter 11
Malignant Melanoma

Chemotherapy Regimens and Cancer Care, by Alan D. Langerak and Luke P. Dreisbach.

Malignant Melanoma

Agent	Dosage
Dartmouth Regimen–dacarbazine (DTIC)/ carmustine (BCNU)/ cisplatin/ tamoxifen	DTIC 220 mg/M² IV days 1-3 BCNU 150 mg/M² IV day 1 of every other cycle Cisplatin 25 mg/M² IV days 1-3 Tamoxifen 20 mg PO daily REF: Chapman et al. J Clin Oncol 1999; 17:2745-2751 PREMEDICATIONS 1. Kytril 1 mg PO/IV 30 minutes before and 12 hours after cisplatin 2. Dexamethasone 20 mg IV 30 minutes before cisplatin OTHER MEDICATIONS 1. Give cisplatin delayed emesis prophylaxis Repeat every 21 days
	Cisplatin—vigorous hydration is required; can be nephrotoxic and ototoxic; can cause peripheral neuropathy; hold or reduce for creatinine > 1.5 Dacarbazine—vesicant—avoid extravasation Carmustine—maximum total dose is 1,440 mg/M²; causes delayed myelosuppression
CDB dacarbazine (DTIC)/ carmustine (BCNU)/cisplatin	–this regimen is the same as the above Dartmouth regimen, with the exception that tamoxifen is not used in CDB –antiemetic regimens and warnings are the same as for the Dartmouth regimen REF: Creagan et al. J Clin Oncol 1999; 17:1884-1890
Paclitaxel/ tamoxifen	Paclitaxel 225 mg/M² IV (over 3 h) day 1 Tamoxifen 40 mg PO daily REF: Nathan et al. Cancer 2000; 88:79-87 PREMEDICATIONS 1. Dexamethasone 20 mg IV 30 minutes before chemotherapy OR Dexamethasone 20 mg PO 6 and 12 hours prior 2. Diphenhydramine 50 mg IV 30 minutes before chemotherapy 3. Cimetidine 300 mg IV 30 minutes before chemotherapy

Continued

Agent	Dosage
	OTHER MEDICATIONS 1. Dexamethasone 4 mg PO BID for 6 doses after paclitaxel (for myalgias) Repeat every 21 days
Vinorelbine/ tamoxifen	Vinorelbine 30 mg/M² IV weekly for 13 wks –after 13 weeks, vinorelbine is given every 2 weeks Tamoxifen 10 mg PO BID daily REF: Feun et al. Cancer 2000; 88:584-588 PREMEDICATIONS 1. Compazine 10 mg PO/IV 30 minutes before chemotherapy Vinorelbine—vesicant; avoid extravasation; can cause peripheral neuropathy
Dacarbazine (DTIC)	DTIC 1000 mg/M² IV day 1 REF: Chapman et al. J Clin Oncol 1999; 17:2745-2751 PREMEDICATIONS 1. Kytril 1 mg PO/IV 30 minutes before and 12 hours after chemotherapy 2. Dexamethasone 20 mg IV 30 minutes before chemotherapy Repeat every 21 days Dacarbazine—vesicant-avoid extravasation
Interferon alfa-2b (IFN)	IFN 20 million units/M² IV days 1-5 weekly X 4 wks –followed by IFN 10 million units/M² SC 3 times weekly X 48 wks REF: Kirkwood et al. J Clin Oncol 1996; 14:7-17 PREMEDICATIONS 1. Tylenol 650 mg PO before each dose This regimen is a one year adjuvant course
High-dose Interleukin-2 (IL-2)	IL-2 600,000-720,000 IU/kg IV (over 15 min) Q8H X 14 doses –repeat above in 6-9 days REF: Atkins et al. J Clin Oncol 1999; 17:2105-2116 PREMEDICATIONS 1. Kytril 1 mg PO/IV 30 minutes before therapy and Q12H during therapy 2. Tylenol 650 mg PO 30 minutes before each dose of IL-2, and Q4H prn

Continued

Agent	Dosage
	3. Cimetidine 800 mg PO/IV daily during IL-2 therapy (given in single or divided doses) Repeat every 6-12 weeks IL-2—may cause capillary leak syndrome with profound hypotension and patients may require vasopressor support and aggressive fluid management. Patients should be cared for in an intensive care setting
Temozolomide	Temozolomide 200 mg/M² PO days 1-5 REF: Middleton et al. J Clin Oncol 2000; 18:158-166 Repeat every 28 days Temozolomide—start at 150 mg/M² and advance dose up to 200 mg/M² as tolerated, based on myelosuppression (adjust dose per package insert); taken for a maximum of 2 years, or until disease progression occurs

Non-Melanoma Skin Cancer

Agent	Dosage			
Cisplatin/ doxorubicin	Doxorubicin	50 mg/M²	IV	day 1
	Cisplatin	75 mg/M²	IV	day 1

REF: Guthrie et al. J Clin Oncol 1990; 8:342-346

PREMEDICATIONS
1. Kytril 1 mg PO/IV 30 minutes before and 12 hours after chemotherapy
2. Dexamethasone 20 mg IV 30 minutes before chemotherapy

OTHER MEDICATIONS
1. Give cisplatin delayed emesis prophylaxis

Repeat every 21 days

Cisplatin—vigorous hydration is required; can be nephrotoxic and ototoxic; can cause peripheral neuropathy; hold or reduce for creatinine > 1.5

Doxorubicin—monitor cumulative dose for cardiac toxicity (not to exceed 550 mg/M² or 450 mg/M² with prior chest radiotherapy); vesicant—avoid extravasation; use 50% for bilirubin 1.5-3.0; use 25% for bilirubin > 3.0

Chapter 12
Sarcoma

- Kaposi's Sarcoma

Sarcoma

<table>
<tr><th>Agent</th><th>Dosage</th></tr>
<tr><td>ADIC doxorubicin/ dacarbazine (DTIC)</td><td>Doxorubicin 60 mg/M² IV day 1
Dacarbazine 250 mg/M² IV (over 1 h) days 1-5

REF: Baker et al. J Clin Oncol 1987; 5:851-861

PREMEDICATIONS
1. Kytril 1 mg PO/IV 30 minutes before and 12 hours after chemotherapy on days 1-5
2. Dexamethasone 10-20 mg IV 30 minutes before chemotherapy on days 1-5

Repeat every 21 days

Doxorubicin—monitor cumulative dose for cardiac toxicity (not to exceed 550 mg/M² or 450 mg/M² with prior chest radiotherapy); vesicant – avoid extravasation; use 50% for bilirubin 1.5-3.0; use 25% for bilirubin > 3.0

Dacarbazine—vesicant—avoid extravasation</td></tr>
<tr><td>CyVADIC cyclophosphamide/ vincristine/ doxorubicin/ dacarbazine (DTIC)</td><td>Cyclophosphamide 500 mg/M² IV day 1
Vincristine 1.4 mg/M² IV day 1
Doxorubicin 50 mg/M² IV day 1
Dacarbazine 400 mg/M² IV days 1-3

REF: Bramwell et al. J Clin Oncol 1994; 12:1137-1149

PREMEDICATIONS
1. Kytril 1 mg PO/IV 30 minutes before and 12 hours after chemotherapy on days 1-3
2. Dexamethasone 10-20 mg IV 30 minutes before chemotherapy on days 1-3

Repeat every 28 days

Doxorubicin—monitor cumulative dose for cardiac toxicity (not to exceed 550 mg/M² or 450 mg/M² with prior chest radiotherapy); vesicant – avoid extravasation; use 50% for bilirubin 1.5-3.0; use 25% for bilirubin > 3.0

Dacarbazine—vesicant—avoid extravasation

Vincristine—vesicant–avoid extravasation; cumulative neurotoxicity—may produce severe constipation; maximum 2 mg per administration</td></tr>
</table>

Agent	Dosage
DI doxorubicin/ ifosfamide/ mesna	Doxorubicin 50 mg/M^2 IV day 1 Ifosfamide 5000 mg/M^2 CIV (over 24 h) day 1 –start after doxorubicin Mesna 600 mg/M^2 IV day 1 bolus before ifosfamide –followed by Mesna 2500 mg/M^2 CIV (over 24 h) day 1 Mesna 1250 mg/M^2 CIV (over 12 h) day 2 REF: Schutte et al. Eur J Cancer 1990; 26:558-561 –there are multiple variations of this regimen PREMEDICATIONS 1. Kytril 1 mg PO/IV 30 minutes before chemotherapy and Q12H for 3 additional doses 2. Dexamethasone 20 mg IV on days 1 and 2 Repeat every 21 days Doxorubicin—monitor cumulative dose for cardiac toxicity (not to exceed 550 mg/M^2 or 450 mg/M^2 with prior chest radiotherapy); vesicant—avoid extravasation; use 50% for bilirubin 1.5-3.0; use 25% for bilirubin > 3.0 Ifosfamide—adequate hydration is necessary to prevent nephrotoxicity
MAID mesna/ doxorubicin/ ifosfamide/ dacarbazine (DTIC)	Mesna 2500 $mg/M^2/d$ CIV (X 96 h) days 1-4 Doxorubicin 20 $mg/M^2/d$ CIV (X 72 h) days 1-3 Ifosfamide 2500 $mg/M^2/d$ CIV (X 72 h) days 1-3 Dacarbazine 300 $mg/M^2/d$ CIV (X 72 h) days 1-3 REF: Elias et al. J Clin Oncol 1989; 7:1208-1216 PREMEDICATIONS 1. Kytril 1 mg PO/IV 30 minutes before chemotherapy on day 1 then Q12H for 6 additional doses 2. Dexamethasone 20 mg IV days 1-3 Repeat every 21-28 days Doxorubicin—monitor cumulative dose for cardiac toxicity; vesicant—avoid extravasation; can give larger cumulative doses than "standard" because less cardiotoxic by continuous infusion; use 50% for bilirubin 1.5-3.0; use 25% for bilirubin > 3.0 Dacarbazine—vesicant—avoid extravasation Ifosfamide—adequate hydration is necessary to prevent nephrotoxicity

Agent	Dosage
Doxorubicin	Doxorubicin 75 mg/M² IV day 1 REF: Santoro, et al. J Clin Oncol 1995; 13:1537-1545 PREMEDICATIONS 1. Kytril 1 mg PO/IV 30 minutes before chemotherapy 2. Dexamethasone 20 mg IV 30 minutes before chemotherapy Repeat every 21 days Doxorubicin—monitor cumulative dose for cardiac toxicity (not to exceed 550 mg/M² or 450 mg/M² with prior chest radiotherapy); vesicant—avoid extravasation; use 50% for bilirubin 1.5-3.0; use 25% for bilirubin > 3.0
Ifosfamide/ mesna	Ifosfamide 5000 mg/M² CIV (X 24 h) day 1 Mesna 400 mg/M² IV Q4H X 9 doses REF: Bramwell et al. Eur J Cancer Clin Oncol 1987; 23:311-321 PREMEDICATIONS 1. Kytril 1 mg PO/IV 30 minutes before chemotherapy 2. Dexamethasone 20 mg IV 30 minutes before chemotherapy Repeat every 21 days Ifosfamide—adequate hydration is necessary to prevent nephrotoxicity

Kaposi's Sarcoma

Agent	Dosage
ABV doxorubicin/ bleomycin/ vincristine	Doxorubicin 10 mg/M² IV day 1 Bleomycin 15 units IV day 1 Vincristine 1 mg IV day 1 REF: Gill et al. J Clin Oncol 1996; 14:2353-2364 PREMEDICATIONS 1. Compazine 10 mg PO/IV 30 minutes before chemotherapy 2. Dexamethasone 10 mg IV 30 minutes before chemotherapy Repeat every 14 days Vincristine—vesicant–avoid extravasation; cumulative neurotoxicity—may produce severe constipation; maximum 2 mg per administration Doxorubicin—monitor cumulative dose for cardiac toxicity (not to exceed 550 mg/M² or 450 mg/M² with prior chest radiotherapy); vesicant—avoid extravasation; use 50% for bilirubin 1.5-3.0; use 25% for bilirubin > 3.0 Bleomycin—give test dose of 1-2 units because of possible acute pulmonary, anaphylactoid, or severe febrile reactions; must dose adjust for renal insufficiency; total lifetime dose should not exceed 400 units; avoid high FiO_2 as it can exacerbate pulmonary toxicity
Liposomal daunorubicin (DaunoXome)	DaunoXome 40 mg/M² IV (over 1 h) day 1 REF: Gill et al. J Clin Oncol 1996; 14:2353-2364 PREMEDICATIONS 1. Kytril 1 mg PO/IV 30 minutes before and 12 hours after chemotherapy 2. Dexamethasone 20 mg IV 30 minutes before chemotherapy Repeat every 14 days Daunorubicin—monitor cumulative dose for possible cardiac toxicity; vesicant—avoid extravasation
Liposomal doxorubicin (Doxil)	Doxil 20 mg/M² IV day 1 REF: Northfelt et al. J Clin Oncol 1997; 15:653-659

Continued

Agent	Dosage
	PREMEDICATIONS 1. Kytril 1 mg PO/IV 30 minutes before and 12 hours after chemotherapy 2. Dexamethasone 20 mg IV 30 minutes before chemotherapy Repeat every 21 days Doxorubicin—monitor cumulative dose for cardiac toxicity (not to exceed 550 mg/M² or 450 mg/M² with prior chest radiotherapy); vesicant—avoid extravasation; use 50% for bilirubin 1.5-3.0; use 25% for bilirubin > 3.0
Paclitaxel	Paclitaxel 100 mg/M² IV (over 3 h) day 1 REF: Gill et al. J Clin Oncol 1999; 17:1876-1883 PREMEDICATIONS 1. Dexamethasone 20 mg IV 30 minutes before chemotherapy OR Dexamethasone 20 mg PO 6 and 12 hours prior to chemotherapy 2. Diphenhydramine 50 mg IV 30 minutes before chemotherapy 3. Cimetidine 300 mg IV 30 minutes before chemotherapy OTHER MEDICATIONS 1. Dexamethasone 4 mg PO BID for 8 doses after chemotherapy Repeat every 14 days

Chapter 13
Supportive Care

- Antiemetics and Guidelines
  - Emetogenic Potential
  - Antiemetics
  - Acute Emesis Guidelines
- Management of Neutropenic Fevers
- Side Effect Management
  - Appetite Stimulants
  - Constipation
  - Diarrhea
  - Extravasation
  - Hiccups
  - Hot Flashes
  - Stomatitis/Oral Care
- Pain Control

Supportive Care

Antiemetics and Guidelines

Agent	Emetogenic Potential	Need for Delayed-Emesis Prophylaxis
Emetogenic potential of chemotherapeutic agents		
Asparaginase	low	
Bleomycin	low	
Carboplatin	high	YES
Carmustine (BCNU)	high	YES
Cisplatin	very high	YES
Cladribine	very low	
Cyclophosphamide		
high dose	high	YES
standard dose	moderate	NO
Cytarabine		
high dose	high	
standard dose	moderate	
Dacarbazine (DTIC)	very high	
Dactinomycin	high	
Daunorubicin	moderate	
Docetaxel	very low	
Doxorubicin	moderate	OCCASIONAL
Epirubicin	moderate	
Etoposide		
high dose	high	
standard dose	low	
Fludarabine	very low	
Fluorouracil		
high dose	moderate	
standard dose	low	
Gemcitabine	low	
Idarubicin	moderate	
Ifosfamide		
high dose	high	
standard dose	moderate	
Irinotecan	low	
Mechlorethamine	very high	
Melphalan		
high dose	very high	
standard dose	low	

Continued

Agent	Emetogenic Potential	Need for Delayed-Emesis Prophylaxis
Methotrexate		
high dose	high	
standard dose	low	
Mitomycin C	moderate	
Mitoxantrone	moderate	
Paclitaxel	very low	
Pentostatin	very low	
Rituximab	very low	
Streptozocin	very high	YES
Thiotepa	low	
Topotecan	moderate	
Trastuzumab	very low	
Vinblastine	low	
Vincristine	very low	
Vinorelbine	low	

Very high → > 90% likelihood of nausea and/or vomiting
High → 60-90% likelihood of nausea and/or vomiting
Moderate → 30-60% likelihood of nausea and/or vomiting
Low → 10-30% likelihood of nausea and/or vomiting
Very low → < 10% likelihood of nausea and/or vomiting

Antiemetics

5-HT3 Antagonists

– these agents are typically used only for acute nausea prophylaxis, and not for delayed emesis prophylaxis; they are usually administered only during the period of chemotherapy administration

Agent	Dosage
Dolasetron (Anzemet)	100 mg PO 30 min before chemotherapy OR 100 mg IV 30 min before chemotherapy
Granisetron (Kytril)	1 mg PO 30 min before and 12 hr after chemotherapy OR 2 mg PO 30 min before chemotherapy OR 0.01 mg/kg IV 30 min before chemotherapy (typical dose is 1 mg)
Ondansetron (Zofran)	8 mg PO 30-60 min before and 8 hr after chemotherapy OR 8-24 mg IV 30 min before chemotherapy
Phenothiazines	
Prochlorperazine (Compazine)	10 mg PO Q4-6H 15 mg spansule PO Q8-12H 25 mg rectal suppository Q4-6H 10 mg IV Q4-6H
Thiethylperazine (Torecan)	10 mg PO Q4-6H 2 mg IM Q4-6H
Trimethobenzamide (Tigan)	250 mg PO Q4-6H 200 mg rectal suppository Q4-6H 200 mg IM Q4-6H
Butyrophenones	
Haloperidol (Haldol)	1-3 mg PO/IV Q4-6H
Droperidol	0.5-2 mg IV Q4H
Benzamide	
Metoclopramide (Reglan)	0.5 mg/kg PO Q6H 1-2 mg/kg IV over 20 min Q3-4H
Benzodiazepines	
Lorazepam (Ativan)	1-2 mg PO/IV/IM Q4-6H

Continued

Agent	Dosage
Cannabinoids	
Dronabinol (Marinol)	2.5-10 mg PO Q6H
Corticosteroids	
Acute emesis	
Dexamethasone	10-20 mg IV prior to chemotherapy for up to 5 days 4-8 mg PO Q4H (for up to 4 doses)
Delayed emesis	
Dexamethasone	8 mg PO BID for 2 days, then 4 mg PO BID for 2 days
Other antiemetics	
Promethazine (Phenergan)	25 mg PO/IV/rectal suppository Q4H
Hydroxyzine (Vistaril)	25 mg PO Q6H

Acute emesis guidelines

Moderate, high, and very high likelihood of nausea/vomiting—5-HT3 antagonist (as above) and dexamethasone (as above) ± benzodiazepine
Low likelihood of nausea/vomiting—Compazine
Very low likelihood of nausea/vomiting—Compazine only if needed

Delayed emesis guidelines

Regimen A
1. Dexamethasone 8 mg PO BID for 2 days, then 4 mg PO BID for 2 days
2. Metoclopramide 0.5 mg/kg QID for 4 days

Regimen B
1. Dexamethasone 8 mg PO BID for 2 days, then 4 mg PO BID for 2 days
2. Prochlorperazine spansules 15 mg PO TID or prochlorperazine 10 mg PO Q4-6H for 4 days

May add diphenhydramine (Benadryl) 50 mg PO Q6H if needed
May add lorazepam 0.5-2 mg PO Q6H if needed

Management of Neutropenic Fever

High Risk Patients

Risk Factors

1. Neutropenia expected to last > 7 days
2. Hematologic malignancies
3. Significant comorbid conditions
4. Clinically evident source of infection
5. Unstable patient (hypoxia, hypotension, etc.)
6. Lack of control of underlying malignancy
7. Transplant patients
8. Elderly
9. Failure of outpatient antibiotics

Treatment protocols

A. No Site of Infection Evident

–use monotherapy or combination therapy as indicated by clinical scenario

1. Monotherapy (one of the following)

a. Ceftazidime	1-2 gm IV Q8H
b. Cefepime	1-2 gm IV Q12H
c. Imipenem/cilastatin	500 mg IV Q6H
d. Levofloxacin	500 mg IV Q24H
e. Piperacillin/tazobactam	3.375 gm IV Q6H
f. Ticarcillin/clavulanate	3.1 gm IV Q6H

2. Combination therapy (one of the following combinations)

a. Antipseudomonal beta-lactam (a, b, c, e, or f above) + aminoglycoside

–gentamicin	5-6 mg/kg/d IV Q24H
–tobramycin	5-6 mg/kg/d IV Q24H
–amikacin	15 mg/kg/d IV Q24H

b. Antipseudomonal beta-lactam (a, b, c, e, or f above) + fluoroquinolone

B. Site of Infection Evident

–therapy should be broad-based, but individualized to the most likely organisms causing infection at that site

Continued

C. Indications for the Use of Vancomycin

1. Catheter infection
2. Severe mucositis
3. Blood culture positive for gram positive organism
4. Known colonization with MRSA or other resistant organism

D. Empiric Antifungal and Antiviral Therapies as Indicated

1. Antifungal

a. Fluconazole 400 mg IV/PO Q24H

b. Itraconazole 200-600 mg PO Q24H or in divided doses

c. Amphotericin B 0.5-1.5 mg/kg IV Q24H

–total dose 1-1.5 gm for Yeast, and 2-2.5 gm for Mold

d. ABLC* (Ablecet) 5 mg/kg IV Q24H over 2 hr

e. ABCD* (Amphocil) 2-4 mg/kg IV Q24H

f. L-AMB* (AmBisome) 3-5 mg/kg IV Q24H over 1-2 hr

*ABLC = Amphotericin B Lipid Complex

*ABCD = Amphotericin B Colloidal Dispersion

*L-AMB = Liposomal Amphotericin B

2. Antiviral

a. Acyclovir 5-10 mg/kg IV Q8H over 1 hr

800 mg PO 5X daily for 7-10 days (herpes zoster)

400 mg PO BID (prophylaxis for herpes)

400 mg PO TID for 5 days (recurrent genital herpes)

b. Famciclovir 500 mg PO TID for 7 days (herpes zoster)

250 mg PO BID (prophylaxis for herpes)

125 mg PO BID for 5 days (recurrent genital herpes)

Continued

c. Valacyclovir	1000 mg PO TID for 7 days (herpes zoster)
	500-1000 mg PO QD (prophylaxis for herpes)
	500 mg PO BID for 5 days (recurrent genital herpes)

Low Risk Patients

Risk Factors

1. Neutropenia expected to last < 7 days
2. Solid tumors
3. No significant comorbid conditions
4. No clinically evident source of infection
5. No significant electrolyte abnormalities
6. Liver tests less than 2-3 times upper limit of normal

Treatment protocols

1. Outpatient IV antibiotics with or without oral antibiotics after appropriate period of observation

   A. Monotherapy or Combination therapy as listed above.

   B. IV antibiotics followed by oral therapy

2. Oral therapy

   A. Ciprofloxacin 500-750 mg PO Q12H +/- Amoxicillin/ clavulanate 875 mg PO Q12H

   B. Ciprofloxacin +/- Clindamycin 150-450 mg PO Q6H in penicillin-allergic patients

REF: Rolston. Clin Infect Dis 1999; 29:515-521

Ramphal. Clin Infect Dis 1999; 29:508-514

Feld. Clin Infect Dis 1999; 29:503-507

Side Effect Management

APPETITE STIMULANTS

Dexamethasone	0.75 mg	PO	QID
OR			
Megestrol acetate	800 mg	PO	QD
- many start at 160 mg QD			
REF: Loprinzi et al. J Clin Oncol 1999; 17:3299-3306			

CONSTIPATION

Bisacodyl (Dulcolax)	10-15 mg PO prn
	10 mg supp PR prn
Castor oil	10-15 cc PO prn
Docusate calcium (Surfak)	240 mg PO QD
Docusate/casanthranol (Pericolace)	1-2 capsules PO QHS prn
	15-30 cc PO QHS prn
Docusate sodium (Colace)	100-200 mg PO BID
Glycerin	1 supp PR prn
Lactulose	15-30 cc PO QHS
Magnesium citrate	150-300 cc PO BID prn
Magnesium hydroxide (MOM)	30-60 cc PO BID prn
Methylcellulose (Citrucel)	1 heaping tablespoon in 8 oz. H_2O TID prn
Mineral oil	15-45 cc PO prn
	120 cc enema PR prn
Polycarbophil (FiberCon)	1 gram PO QID prn
Polyethylene glycol (MiraLax)	17 gms (1 heaping tbs) in 8 oz. H_2O QD
Psyllium (Metamucil)	1 teaspoon in liquid, 1 packet in liquid, or 1-2 wafers PO TID prn

Continued

	Senna (Senokot)	2 tabs or 1 teaspoon of granules or 10-15 cc syrup PO QHS prn
		1 supp PR QHS prn
	Sodium bisphosphate (Fleet)	1 enema PR prn
	Sorbitol	30-150 cc PO prn
DIARRHEA		
	Kaolin and pectin (Kaopectate)	15-30 cc PO Q4H prn
	Loperamide (Imodium)	4 mg PO after first loose BM, then 2 mg PO after each loose BM
	–should not to exceed 6 doses per 24 hours	
	Diphenoxylate/atropine (Lomotil)	1-2 tabs PO Q4H prn
	Octreotide	0.05-0.1 mg SQ TID
	–may be helpful for 5-FU induced diarrhea	

EXTRAVASATION

ANTHRACYCLINES

	–dactinomycin, daunorubicin, doxorubicin, epirubicin, idarubicin, ± mitoxantrone
Treatment	1. Application of cold – apply without pressure on and off for 24 hours 2. Topical DMSO – 1.5 cc topically Q6H for 14 days; allow to air dry

NITROGEN MUSTARD AND MITOMYCIN C

Treatment	1. Mix 4 cc 10% sodium thiosulfate with 6 cc of sterile H_20 –inject 2 cc into site for each mg of drug extavasated 2. Topical DMSO – 1.5 cc topically Q6H for 7-14 days; allow to air dry * cisplatin—large extravasations can also be treated in the above manner

VINCA ALKALOIDS

	–vinblastine, vincristine, vinorelbine
Treatment	1. 150 units hyaluronidase reconstituted in 1-3 cc sterile saline –inject into site using original needle if possible 2. Warm pack—apply to site without pressure after above injection

Epipodophyllotoxins		
	–etoposide (VP-16), teniposide (VM-26)	
Treatment	1. treat for large volume extravasations 2. 150 units hyaluronidase reconstituted in 1-3 cc sterile saline –inject into site using original needle if possible 3. Warm pack—apply to site without pressure after above injection	
HICCUPS		
	Amitriptyline	25 mg PO BID
	Baclofen	10 mg PO Q6-8H
	Carbamazepine	200 mg PO QID prn
	Chlorpromazine (Thorazine)	25-50 mg IM
	Lorazepam (Ativan)	0.5-1 mg PO/IV Q6H prn
	Metoclopramide	10-20 mg PO QID prn
	Prochlorperazine (Compazine)	10mg PO Q6-8H
	Simethicone	40-160 mg PO QID prn
HOT FLASHES		
	Bellergal-S	1 PO QD-BID (start QHS)
	Clonidine	0.1 mg patch weekly
	Megestrol (Megace)	20-40 mg PO BID-QID
	Methyldopa	250 mg PO BID
	Venlafaxine	75 mg PO QD
	Vitamin B6	200 mg PO QD
	Vitamin E	800 IU PO QD
HYPERURICEMIA		
	Allopurinol	300-600 mg PO QD 200-400 mg/M^2 IV QD (not to exceed 600 mg QD)

STOMATITIS/ORAL CARE

Chlorhexidine (Peridex)	15 cc swish & spit QID
Sodium bicarbonate	1 tsp baking soda in 500 cc water
	15 cc swish & spit QID
Clotrimazole (Mycelex) troche	1 troche dissolved in mouth 5X daily
"Miracle" mouthwash	Diphenhydramine (12.5 mg/5 cc) 420 cc
	Dexamethasone (500 mcg/ml) 90 cc
	Nystatin suspension (100,000 units/cc) 120cc
	Sterile water 330 cc
	–5 cc swish and swallow QID

–there are many variations of this "recipe"

Viscous lidocaine	5-10 cc swish & spit
Vitamin E	puncture capsule and apply to oral lesions
Zilactin gel	Apply to lesions QID
Kaopectate	10 cc swish & swallow prn
Magnesium hydroxide (MOM)	10 cc swish & swallow prn
Maalox	10 cc swish & swallow prn

PAIN CONTROL

NARCOTICS FOR SEVERE PAIN

Name	Starting Dose	Route	Frequency	Dosage Forms
Morphine sulfate	1-2 mg	IV/SQ/IM	Q2-4H prn	0.5, 1 mg/ml
SR	15-30 mg	PO	Q8-12H	15,30,60,100 mg
IR	15-30 mg	PO	Q2-4H prn	15,30 mg
IR-solution	15-30 mg	PO	Q2-4H prn	10,20 mg/5 ml 20 mg/ml
CR	15-30 mg	PO	Q8-12H	15,30,60, 100,200 mg
Suppository	10-30 mg	Rectally	Q4-6H prn	5,10,20,30 mg
Oxycodone	5 mg	PO	Q4-6H prn	5 mg
Solution	5 mg	PO	Q4-6H prn	5 mg/5 ml, 20 mg/ml
SR	10-20 mg	PO	Q8-12H	10,20,40,80 mg
with APAP	1–2 tabs	PO	Q4-6H prn	5 mg/325 mg (Percocet) 5 mg/500 mg (Tylox)
Hydromorphone	2 mg	PO	Q4-6H prn	2,4,8 mg
Oral Liquid	2.5 mg	PO	Q3-6H prn	5 mg/5 ml
Suppository	3 mg	Rectally	Q6-8H prn	3 mg
Injectable	1-2 mg	IV/SQ/IM	Q4-6H prn	1,2,4 mg/ml
Methadone	2.5-5 mg	PO	Q4-6H prn	5,10 mg
Injectable	2.5-5 mg	SQ/IM	Q4-6H prn	10 mg/ml
Meperidine	50 mg	PO	Q3-4H prn	50, 100 mg
Syrup	50 mg	PO	Q3-4H prn	50 mg/5 ml
Injectable	50 mg	IV/SQ/IM	Q3-4H prn	25,50,75, 100 mg/ml
Fentanyl	25 mcg/hr	Transdermal	Q72H	25,50,75, 100 mcg/hr
Lozenge	200 mcg	PO	Q4-6H prn	200,300,400 mcg

Narcotics for mild-moderate pain

Name	Starting Dose	Route	Frequency	Dosage Forms
Codeine	15-30 mg	PO	Q4-6H prn	15, 30, 60 mg
Injectable	15-30 mg	IV/SQ/IM	Q4-6H prn	30, 60 mg/ml
with APAP	1-2 tabs	PO	Q4-6H prn	15 mg/300 mg (Tylenol #2)
with APAP	1-2 tabs	PO	Q4-6H prn	30 mg/300 mg (Tylenol #3)
with APAP	1-2 tabs	PO	Q4-6H prn	60 mg/300 mg (Tylenol #4)
Hydrocodone with APAP	1-2 tabs	PO	Q4-6H prn	2.5 mg/500 mg (Lortab)
with APAP	1-2 tabs	PO	Q4-6H prn	5 mg/500 mg (Vicodin)
				5 mg/500 mg (Lortab)
with APAP	1 tab	PO	Q4-6H prn	7.5 mg/500 mg (Lortab)
				7.5 mg/750 mg (Vicodin ES)
Propoxyphene	1 tab	PO	Q4-6H prn	65,100 mg (Darvon)
with APAP	1 tab	PO	Q4-6H prn	100 mg/650 mg (Darvocet N-100)
NSAIDS				
Ibuprofen	400 mg	PO	Q6-8H prn	200,300,400, 600,800 mg
Suspension	400 mg	PO	Q6-8H prn	100 mg/5 ml
Naproxen	220-500 mg	PO	Q12H prn	220,250,275, 375,500,550 mg
Suspension	250 mg	PO	Q12H prn	125 mg/5 ml
Salsalate	750-1000 mg	PO	Q12H prn	500,750 mg
Oxaprozin	600 mg	PO	Q24H prn	600 mg (Daypro)
Sulindac	150 mg	PO	Q12H prn	150,200 mg (Clinoril)

Continued

Nabumetone	1000 mg	PO	Q24H prn	500,750 mg (Relafen)
Piroxicam	10 mg	PO	Q24H prn	10,20 mg (Feldene)
Celecoxib	100 mg	PO	Q12H	100,200 mg (Celebrex)
Rofecoxib	12.5 mg	PO	Q24H	12.5,25 mg (Vioxx)

OTHER ANALGESICS

A. Antidepressants

Amitriptyline	10-25 mg	PO	QHS	10,25,50,75, 100 mg (Elavil)
Desipramine	10-25 mg	PO	QHS	10,25, 50 mg (Norpramin)
Paroxetine	20 mg	PO	QD	20,40 mg (Paxil)
Sertraline	25 mg	PO	QD	50,100 mg (Zoloft)
Citalopram	20 mg	PO	QD	20,40 mg (Celexa)

B. Anticonvulsants

Gabapentin	100 mg	PO	Q8H	100,300,400 mg (Neurontin)
Carbamazepine	100 mg	PO	BID	100,200,400 mg (Tegretol) 100 mg/5cc suspension

C. Miscellaneous

Corticosteroids

Dexamethasone	4 mg	PO	Q6H	0.75,1,2,4 mg (Decadron)

–taper dose to lowest dose which is effective

Stimulants

Methylphenidate	5 mg	PO	BID	5,10,20 mg (Ritalin)
Dextroamphetamine	5 mg	PO	BID	5,10,15 mg (Dexedrine)

Chapter 14
Hematology Drugs

- Anticoagulation
- Aplastic Anemia
- Coagulation Factor Replacement Therapy
- Hematopoietic Growth Factors
- Immune Thrombocytopenic Purpura
- Iron Replacement
- Thrombocytosis

Chemotherapy Regimens and Cancer Care, by Alan D. Langerak and Luke P. Dreisbach.

Hematology Drugs

Anticoagulation

Warfarin	–adjust dosage to maintain INR of 2-3 (3-4.5 for prosthetic valves)			
Unfractionated Heparin	–loading dose of 80 mg/kg, followed by 18 mg/kg/hr; adjust dose to maintain a therapeutic PTT REF: Raschke et al. Ann Intern Med 1993; 119:874-881			
Low-Molecular Weight Heparin				
Prophylaxis	Enoxaparin (Lovenox)	30-60 mg	SQ	BID
	Dalteparin (Fragmin)	2500-5000 units	SQ	QD
	Ardeparin (Normiflo)	50 units/kg	SQ	BID
Treatment	Enoxaparin (Lovenox)	1 mg/kg	SQ	Q12H
	Dalteparin (Frågmin)	100 units/kg	SQ	Q12H
Heparin Reversal				
Unfractionated Heparin	Protamine sulfate	1 mg per 100 units (if PTT prolonged 2-4 hours later, give 1/2 of initial dose)		
Dalteparin	Protamine sulfate	1 mg per 100 units		
Enoxaparin	Protamine sulfate	1 mg per mg of enoxaparin		
Heparanoids	–can be used in heparin-induced thrombocytopenia			
	Danaparoid (Orgaran)	1250 units IV load, followed by 1250 units	SQ	Q12H
	REF: de Valk et al. Ann Intern Med 1995; 123:1-9			
Direct Thrombin Inhibitors				
Prophylaxis	–can be used in heparin-induced thrombocytopenia			
	Lepirudin (Refludan)	0.1 mg/kg/hr		

Continued

Lung Cancer
Malignant Melanoma
Sarcoma
Supportive Care
Hematology Basics
Chemotherapeutic Drug Toxicities
Drug Costs

Treatment	Lepirudin (Refludan)	0.4 mg/kg IV bolus, followed by 0.15 mg/kg/hr IV infusion to maintain a PTT of 1.5-3 times normal		
	REF: Greinacher et al. Circulation 1999; 100:587-593			
Antiplatelet Agents				
	Ticlopidine	250 mg	PO	TID
	Clopidogrel	75 mg	PO	QD
Thrombolytics				
	Streptokinase	1.5 million units IV over 1 hour		
	Alteplase	100 mg IV: give 60 mg IV during first hour (6-10 mg IV bolus over 1-2 minutes), followed by 20 mg IV during 2nd hour and 20 mg IV during 3rd hour		
	Anistreplase	30 units IV over 2-5 minutes		
	Reteplase	10 unit IV bolus, followed by 10 units IV bolus 30 minutes later		
Antifibrinolytics				
	Aminocaproic Acid (Amicar)	IV	5 gram bolus, followed by 500-1000 mg/hr	
		PO	5 gram bolus, followed by 1-2 grams PO Q1-2HPRN	
	Tranexamic Acid	IV	10 mg/kg	Q6-8H
		PO	25 mg/kg	Q6-8H

Aplastic Anemia

ATG Protocol

ATG Test Dose

ATG 1:1000 dilution in normal saline 0.1 cc intradermally
Control saline 0.1 cc intradermally

Premedication for ATG

Tylenol 650 mg PO 30 minutes before ATG
Benadryl 50 mg PO/IV 30 minutes before ATG
Hydrocortisone 50 mg IV 30 minutes before ATG

ATG Dosing

ATG 40 mg/kg in 1 liter normal saline IV over 8-12 hours QD days 1-4

Concomitant Medications

Prednisone 100 mg/M2 PO QD X 7 days; start with ATG → taper over 7 days if no serum sickness
Cyclosporine 5 mg/kg/d divided BID; taper by 1 mg/kg/month, as tolerated
–start at 4 mg/kg/d if age > 50

Other Therapies to Consider

Hematopoietic growth factors
Cyclosporine alone → fewer remissions than combination with ATG
Androgens (such as Danazol—see dosing in ITP section)—can take 3 or more months to show effect
Other immunosuppressants, such as azathioprine or cyclophosphamide

Coagulation Factor Replacement Therapy

Fresh frozen plasma (FFP)

–used in the absence of a specific factor concentrate, for massive transfusion, to correct warfarin effect, and in TTP with plasma exchange

–FFP dosage is 8-10 ml/kg of body weight (each unit of FFP is approximately 200-280 cc)

Cryoprecipitate

–can be used to replace Factor VIII, Factor XIII, fibrinogen, and von Willebrand factor

–typical dosing is 2-4 units/kg of body weight

Factor VIII

–1 unit/kg will raise plasma factor VIII level by 2%

–purity is based on number of factor VIII units per mg of contaminating protein)

$$\text{Replacement dose for Factor VIII} = \frac{(\text{desired concentration} - \text{current level}) \times \text{wt (kg)}}{2}$$

Low purity (< 50 factor VIII units/mg protein)
–Cryoprecipitate
Intermediate purity (1-10 factor VIII units/mg protein)
–Humate-P (also contains high molecular weight multimers of von Willebrand factor)
   –vials contain average of 500, 1000, or 2000 Ristocetin cofactor units per vial
High purity (50-1000 factor VIII units/mg protein)
–Alphanate
–Koate-HP
Very high purity (3000 factor VIII units/mg protein)
–Monoclate-P—average of 250, 500, or 1000 factor VIII units/vial
–Hemofil-M
Recombinant
–Helixate—average of 250, 500, or 1000 factor VIII units/vial
–Bioclate—average of 250, 500, or 1000 factor VIII units/vial
–Kogenate—average of 250, 500, or 1000 factor VIII units/vial
–Recombinate—average of 250, 500, or 1000 factor VIII units/vial

Continued

For patients with factor VIII inhibitors
–FEIBA VH IMMUNO
–give 50-100 "IMMUNO" units/kg body weight; repeated at 6-12 hour intervals

–Proplex T (Factor IX Complex)—used for factor VIII inhibitors, and factor VII or IX deficiency

–Factor VIII inhibitor dose = 75 factor IX units/kg

Factor IX

Replacement dose for
Factor IX = (desired concentration – current level) X wt (kg)

–multiply this value by 1.2 when using recombinant factor IX

Low purity (< 50 factor IX units/mg protein)
–Proplex T (Factor IX Complex)—used for factor VIII inhibitors, and factor VII or IX deficiency
–Factor IX replacement dose = desired increase X wt (kg)
–Factor VII replacement dose = desired increase in factor VII level X wt (kg) X 0.5
High purity (> 160 factor IX units/mg protein)
–Mononine—average of 250, 500, or 1000 factor IX units/vial
Recombinant
–BeneFix—average of 250, 500, or 1000 factor IX units/vial

DDAVP

IV dose = 0.3 μg/kg over 30 minutes
Nasal dose less than 50 kg → 1 spray (150 μg)
more than 50 kg → 1 spray to each nostril (150 μg each)

Hematopoietic Growth Factors

Erythropoietin (Procrit)

- starting dose is 150 units/kg SQ TIW; can increase dose to 300 units/kg SQ TIW if no response
- many recommend once weekly dosing, using 20-40,000 units SQ once weekly

Filgrastim (Neupogen)—G-CSF

- 5 μg/kg/d IV or SQ

Sargramostim (Leukine)—GM-CSF

- 250 μg/M^2/d IV or SQ

Oprelvekin (Neumega)

- 50 μg/kg/d SQ

Immune Thrombocytopenic Purpura (ITP)

Prednisone

dosed at 1-2 mg/kg by mouth daily; dose is slowly tapered over several weeks to prevent recurrence of thrombocytopenia

REF: Thompson et al. Arch Intern Med 1972; 130:730-734

Dexamethasone

40 mg PO QD days 1-4 every 28 days

REF: Andersen: NEJM 1994; 330:1560-1564

IVIG

1 gm/kg/d IV for 2 days (if thrombocytopenia is less severe, can spread total 2 gm/kg dose over 5 days)

REF: Blanchette et al. Semin Hematol 1992; 29(Suppl 2):72-82

WinRho

25-50 μg/kg IV as initial dosage; some clinicians have given as much as 80 μg/kg (typical adult dose is approximately 2 mg)

REF: Scaradavou et al. Blood 1997; 89:2689-2700

Danazol

200 mg PO QID; responses can take 3-6 months

REF: Ahn et al. NEJM 1983; 308:1396-1399

Vincristine

1-2 mg IV weekly; no more than 4 to 6 doses because of neuropathy; occasional complete responses

REF: Ahn et al. NEJM 1974; 291:376-380

Cyclophosphamide

2 mg/kg PO QD; taper dose as tolerated (increased risk of second malignancies; increased fluid intake to prevent hemorrhagic cystitis)

REF: Pizzuto et al. Blood 1984; 64:1179-1183

Iron Replacement and Chelation

Oral formulations

Ferrous gluconate	(Fergon)	320-640 mg TID
Ferrous sulfate		325 mg tablet TID
	(Feosol)	220 mg/5 cc 5-10 cc TID
Ferrous polysaccharide	(Niferex)	150 mg capsule BID
		100 mg/5cc BID-TID

Intravenous iron

Formula to calculate amount of IV iron

Iron dose (mg) = [(Normal Hb – Patient Hb) X weight (lbs)] + 1000 mg (males) or 600 mg (females)

Iron dextran (InFed) comes as 50 mg/ml

- Premedicate with Diphenhydramine 50 mg PO/IV 30 minutes before iron
- Premedicate with Tylenol 650 mg PO 30 minutes before iron
- Administer test dose of iron 25 mg IV; wait at least 30 minutes; if no reaction →
- Administer remainder of total iron dose in 1 liter normal saline over 4-5 hours
- Tylenol 650 mg PO Q6H for 2 doses after conclusion of iron infusion

Iron chelation therapy

Desferroxamine 40-50 mg/kg SQ over 8-12 hours daily for 5 days weekly

–continue until ferritin is < 50

Thrombocytosis

Hydroxyurea

500-2000 mg by mouth daily (in divided doses) to control platelet count

REF: Lofvenberg et al. Eur J Haematol 1988; 41:375-381

Anagrelide

starting dose is 0.5-1 mg by mouth QID to control platelet count

REF: Anagrelide Study Group: Am J Med 1992; 92:69-76

Chapter 15
Chemotherapeutic Drug Toxicities and Mechanisms of Action

Chemotherapeutic Drug Toxicities

Mechanisms of Action

The following is a list of the most common side effects of each chemotherapeutic agent, along with the proposed mechanism of action for that drug. Please refer to the PDR for a complete toxicity profile. The generally recognized dose-limiting toxicity (DLT) of each drug is underlined.

Aldesleukin (IL-2)	–biologic agent **–capillary leak syndrome (pulmonary edema)–DLT for high-dose administration** **–malaise, myalgias, fatigue–DLT for low-dose administration** –bone marrow suppression –nausea and vomiting –mucocutaneous effects (stomatitis, mucositis) –cardiovascular effects (arrhythmias, hypotension) –anorexia –mental status changes (confusion, lethargy, psychosis) –renal impairment –fever
Altretamine (hexamethyl-melamine)	–alkylating agent **–nausea and vomiting** –bone marrow suppression –diarrhea, abdominal cramps –mucocutaneous effects (stomatitis, mucositis) –neuropathies –mental status changes
Amifostine	–cytoprotectant; free radical scavenger –nausea and vomiting –somnolence –transient hypotension
Aminoglute-thimide	–aromatase inhibitor **–adrenal insufficiency** –mucocutaneous effects—morbilliform rash –lethargy
Anagrelide	–inhibitor of platelet aggregation which causes thrombocytopenia **–cardiovascular effects (CHF, edema, palpitations)** –anemia –nausea and vomiting –headache
Anastrazole	–nonsteroidal aromatase inhibitor –nausea and vomiting –bowel changes (diarrhea or constipation) –headache –peripheral edema –hot flashes

Arsenic trioxide	–believed to induce apoptosis –LFT elevations –renal insufficiency –fatigue –hyperglycemia –skin rash –hypokalemia –peripheral neuropathy –high frequency hearing loss
Asparaginase	–enzyme that inhibits protein synthesis **–anaphylaxis** –hepatotoxicity –CNS effects (lethargy, confusion, somnolence, depression) –coagulopathy –pancreatitis

Bicalutamide	–nonsteroidal antiandrogen
	–endocrine effects
	  –hot flashes
	  –decreased libido
	  –depression
	  –weight gain
	–constipation

Bleomycin	–antitumor antibiotic that causes DNA strand breakage **–dose-related pneumonitis** –mucocutaneous effects (stomatitis, mucositis) –acute pulmonary edema –fever in 50% –hyperpigmentation (can rarely be DLT)
Busulfan	–alkylating agent **–bone marrow suppression—can have prolonged nadir** –ovarian suppression –seizures –hepatic veno-occlusive disease (VOD), particularly at BMT doses –interstitial pulmonary fibrosis –hyperpigmentation (particularly skin creases and nail beds)
Capecitabine	–converted to 5-FU preferentially by tumor cells; pyrimidine analogue; antimetabolite; inhibits thymidylate synthase –mucocutaneous effects (stomatitis, mucositis) –diarrhea –bone marrow suppression –nausea and vomiting –palmar-plantar erythrodysethesias (hand-foot syndrome) –fatigue

Carboplatin	–atypical alkylating agent leading to DNA strand breakage during replication **–bone marrow suppression—particularly thrombocytopenia** –nausea and vomiting –liver function test abnormalities –uncommon neurotoxicity, ototoxicity
Carmustine (BCNU)	–alkylating agent (cell cycle-independent mechanism) **–bone marrow suppression—delayed with a nadir of 3-5 weeks** –nausea and vomiting—can be severe and prolonged –facial flushing –interstitial lung disease (dose independent)
Chlorambucil	–alkylating agent (cell cycle-independent) **–bone marrow suppression** –nausea and vomiting –CNS stimulation (uncommon)
Cisplatin	–atypical alkylating agent leading to DNA strand breakage during replication **–nephrotoxicity—DLT for single dose** –peripheral neuropathy—DLT for multiple doses –bone marrow suppression –nausea and vomiting—can be severe and prolonged –ototoxicity –hypomagnesemia
Cladribine (2-CdA)	–purine analogue; antimetabolite –bone marrow suppression –fever in 50% (probably due to tumor lysis) –rash in 50% –immunosuppression (with profound T-cell lymphopenia)
Cyclophos-phamide	–alkylating agent (cell cycle independent) **–bone marrow suppression** –anorexia, nausea and vomiting –alopecia –hemorrhagic cystitis
Cyclosporine	–immunosuppressant **–nephrotoxicity** –hirsutism –hepatotoxicity –tremor –anxiety –hypertension

Lung Cancer
Malignant Melanoma
Sarcoma
Supportive Care
Hematology Basics
Chemo-therapeutic Drug Toxicities
Drug Costs

Cytarabine (Ara-C)	–antimetabolite which is S-phase specific during DNA replication –bone marrow suppression –nausea and vomiting –cerebellar toxicity (particularly at high doses) –conjunctivitis (at high doses) –hepatotoxicity –mucocutaneous effects (stomatitis, mucositis, diarrhea)
Dacarbazine (DTIC)	–atypical alkylating agent, noncell cycle dependent **–bone marrow suppression** –nausea and vomiting –vesicant if extravasated –flu-like syndrome –fever
Dactinomycin	– antitumor antibiotic; inhibits transcription by complexing with DNA **–bone marrow suppression** –nausea and vomiting –erythema –hyperpigmentation –mucocutaneous effects (mucositis, stomatitis, diarrhea) –vesicant if extravasated –immunosuppression
Daunorubicin	–anthracycline antitumor antibiotic; DNA intercalating agent **–bone marrow suppression** –nausea and vomiting—mild to moderate –mucocutaneous effects (mucositis, stomatitis, diarrhea) –vesicant if extravasated –cardiotoxicity (550 mg/M^2) **–Liposomal daunorubicin**: there is significantly less bone marrow suppression, nausea and vomiting, stomatitis, and cardiotoxicity
Dexamethasone	–corticosteroid –leukocytosis –nausea and vomiting –anorexia or increased appetite –CNS effects (psychosis, confusion) –fluid retention –hyperglycemia –osteoporosis
Dexrazoxane	–iron chelating agent (cardioprotectant) **–leukopenia and thrombocytopenia** –nausea and vomiting –elevated liver function tests –hypotension

Diethylstil besterol (DES)	–synthetic steroidal pro-estrogen hormone –nausea and vomiting –cramps –elevated liver function tests –headache –thromboembolic events –weight gain –rash
Docetaxel	–semisynthetic taxane; stabilizes tubulin polymers leading to death of mitotic cells **–bone marrow suppression** –nausea and vomiting –mucocutaneous effects (mucositis, stomatitis, diarrhea) –hypersensitivity reactions –fluid retention syndrome –fatigue –myalgias –alopecia (universal)
Doxorubicin	– anthracycline antitumor antibiotic – DNA intercalating agent **–bone marrow suppression** –nausea and vomiting –mucocutaneous effects (mucositis, stomatitis) –cardiotoxicity (550 mg/M^2) –vesicant if extravasated –rash and hyperpigmentation –alopecia (universal)
Liposomal doxorubicin—bone marrow suppression; significantly less stomatitis, exstravasation necrosis, and cardiotoxicity	
Epirubicin	–anthracycline antitumor antibiotic—DNA intercalating agent **–bone marrow suppression** –nausea and vomiting –mucocutaneous effects (mucositis, stomatitis) –cardiotoxicity (1000 mg/M^2) –vesicant if extravasated –rash and hyperpigmentation –alopecia
Erythropoietin	–hormonal stimulant of red blood cell production –erythrocytosis (with excessive dosage) –flushing
Estramustine	–inhibitor of microtubules **–nausea and vomiting** –headache –edema –impotence –gynecomastia –increases thromboembolic risk

Etoposide (VP-16)	–plant alkaloid, topoisomerase II inhibitor **–bone marrow suppression** –nausea and vomiting –mucocutaneous effects (mucositis, stomatitis)—increased at higher doses –chemical phlebitis common –hypotension with rapid administration –hypersensitivity reactions –secondary leukemia
Exemestane	–aromatase inhibitor –nausea and vomiting –headache –peripheral edema –hot flashes
Filgrastim (G-CSF)	–hematopoietic growth factor –bone pain –low-grade fever –myalgias, arthralgias –leukocytosis (with excessive dosing) –capillary leak syndrome
Fludarabine	–purine analogue; antimetabolite; partially cell cycle specific **–bone marrow suppression** –nausea and vomiting –mucocutaneous effects (mucositis, stomatitis)—increased at higher doses –CNS toxicity—cortical blindness, confusion, coma, somnolence –interstitial pneumonitis –immunosuppression
5-Fluorouracil (5-FU)	–pyrimidine analogue; antimetabolite; inhibits thymidylate synthase **–mucocutaneous effects (diarrhea, mucositis, stomatitis)** –bone marrow suppression –nausea and vomiting –palmar-plantar erythrodysethesias (hand-foot syndrome) –cardiotoxicity (ischemia, arrhythmias) –acute cerebellar syndrome

Fluoxy-mesterone	–synthetic steroidal androgen
	–androgenic effects predominate
	  –hirsuitism
	  –amenorrhea
	  –hoarseness
	  –acne
	  –increased libido
	  –gynecomastia
	–cholestatic jaundice
	–polycythemia

Flutamide	–nonsteroidal antiandrogen –endocrine effects –hot flashes –decreased libido –gynecomastia –impotence –galactorrhea –diarrhea –nausea and vomiting –myalgias –elevated liver function tests
Gemcitabine	– nucleoside analogue; antimetabolite; S-phase specific cytotoxicity **–bone marrow suppression—most commonly thrombocytopenia** –nausea and vomiting –fever during administration –elevated transaminases –rash
Gemtuzumab zoqamicin	–monoclonal antibody against CD33 with calicheamicin (antitumor antibiotic) –fevers and chills –hypotension –grade IV neutropenia and thrombocytopenia –LFT elevations
Goserelin	– LHRH agonist –endocrine effects –hot flashes –decreased libido –gynecomastia –impotence –nausea and vomiting (uncommon) –transient increase in bone pain
Hydroxyurea	–antimetabolite; inhibits ribonucleotide reductase; cell cycle specific **–bone marrow suppression** –nausea and vomiting (uncommon at standard doses) –maculopapular rash –skin ulceration –megaloblastosis (elevated MCV)
Idarubicin	–anthracycline antitumor antibiotic; DNA intercalating agent **–bone marrow suppression** –nausea and vomiting—mild to moderate –mucocutaneous effects (mucositis, stomatitis, diarrhea) –vesicant if extravasated –cardiotoxicity (150 mg/M^2) –elevated liver function tests

Ifosfamide	– alkylating agent; noncell cycle specific **–bone marrow suppression** **–hemorrhagic cystitis (need Mesna uroprotection)** –nausea and vomiting—mild to moderate –mucocutaneous effects (mucositis, stomatitis, diarrhea) –CNS toxicity—lethargy, stupor, coma, seizures
Interferon	–biologic agent **–flu-like symptoms–malaise, myalgias, fatigue, fever** –nausea and vomiting—mild –anorexia –bone marrow suppression –mucocutaneous effects (stomatitis, mucositis) –cardiovascular effects (arrhythmias, hypotension) –mental status changes (confusion, lethargy, psychosis) –renal impairment (proteinuria) –elevation in transaminase levels
Irinotecan	–semisynthetic camptothecin; topoisomerase I inhibitor **–bone marrow suppression** –diarrhea –nausea and vomiting –flushing –rash –alopecia
Leucovorin (folinic acid)	– enzyme cofactor for thymidylate synthase; rescues from methotrexate toxicity; potentiates cytotoxicity of fluoro–pyrimidines –occasional nausea –skin rash –headache –rare allergic reactions

Leuprolide	–LHRH agonist
	–endocrine effects
	  –hot flashes
	  –decreased libido
	  –gynecomastia (3%)
	  –breast tenderness
	  –impotence (2%)
	–nausea and vomiting (uncommon)
	–transient increase in bone pain
	–peripheral edema
	–dizziness, headache

Levamisole	–immune potentiating effects –nausea and vomiting –diarrhea –anorexia –rash (23%) –alopecia (22%) –rare agranulocytosis (more often in women)

Lomustine (CCNU)	–nitrosourea alkylating agent; cell cycle independent **–bone marrow suppression (delayed, prolonged, and cumulative)** –nausea and vomiting –pulmonary fibrosis –neurologic toxicity – confusion, lethargy, ataxia
Mechloreth-amine (nitrogen mustard)	–alkylating agent; cell cycle independent –bone marrow suppression –vesicant if extravasated –severe nausea and vomiting –impaired spermatogenesis and amenorrhea –maculopapular skin rash –secondary leukemias
Megestrol acetate	–steroidal progestational agent –nausea and vomiting –headache –peripheral edema –hot flashes –thrombophlebitis –increased appetite with weight gain –hypercalcemia
Melphalan	–alkylating agent; cell cycle independent **–bone marrow suppression** –nausea and vomiting (more frequent with large, single oral doses) –pulmonary fibrosis –vasculitis –secondary leukemia
6-Mercapto-purine (6-MP)	– purine analogue antimetabolite; predominantly S-phase specific **–bone marrow suppression** –nausea and vomiting—mild to moderate –mucocutaneous effects (mucositis, stomatitis, diarrhea) –hepatotoxicity –dry scaling rash –fever –eosinophilia
Mesna	–thiol uroprotectant (binds and inactivates toxic metabolite acrolein) –nausea and vomiting –rash –headache –fatigue and lethargy

Lung Cancer
Malignant Melanoma
Sarcoma
Supportive Care
Hematology Basics
Chemo-therapeutic Drug Toxicities
Drug Costs

Methotrexate	–antifolate antimetabolite; cell cycle dependent **–bone marrow suppression** –nausea and vomiting—mild to moderate –mucocutaneous effects (mucositis, stomatitis, diarrhea) –hepatotoxicity—more common in high-dose therapy –CNS toxicity—dizziness, malaise, blurred vision, encephalopathy –nephrotoxicity—including acute renal failure, particularly at high doses
Mitomycin C	–antitumor antibiotic; inhibits RNA and DNA synthesis **–bone marrow suppression** –nausea and vomiting—mild to moderate –mucocutaneous effects (mucositis, stomatitis, diarrhea) –vesicant if extravasated –nephrotoxicity –veno-occlusive disease (VOD) of the liver –hemolytic-uremic syndrome
Mitotane (o,p-DDD)	–adrenocortical cytotoxin **–nausea and vomiting** –CNS toxicity—lethargy, vertigo, sedation, dizziness –adrenal insufficiency—must use replacement doses of mineralocorticoids and glucocorticoids –diarrhea –fever –wheezing –flushing
Mitoxantrone	–anthracycline antitumor antibiotic; DNA intercalating agent **–bone marrow suppression** –nausea and vomiting—mild to moderate –mucocutaneous effects (mucositis, stomatitis, diarrhea) –cardiotoxicity (160 mg/M^2) –elevated liver function tests
Octreotide	–synthetic peptide analogue of somatostatin **–abdominal pain, nausea, vomiting, diarrhea** –local injection site reactions –cholelithiasis –sweating, flushing –hyperglycemia (many patients will require insulin therapy)
Oprelvekin (IL-11, Neumega)	–stimulation of megakaryoctye proliferation –fluid retention –constitutional symptoms—headache, fever, malaise –dyspnea –rash –diarrhea –pleural effusions –anemia

Oxaliplatin	– alkylating agent; causes DNA cross-linking **–peripheral neuropathy (cumulative)—often reversible with cessation of therapy** –mild bone marrow suppression –nausea and vomiting (which may be severe)
Paclitaxel	–natural taxane; inhibits depolymerization of tubulin in mitotic spindle apparatus **–bone marrow suppression** –nausea and vomiting—mild –mucocutaneous effects (mucositis, stomatitis, diarrhea) –hypersensitivity reactions –peripheral neuropathy –myalgias, arthralgias –mild vesicant
Pamidronate	–organic bisphosphonate; inhibits bone resorption by osteoclasts –hypotension –syncope –tachycardia –hypocalcemia, hypokalemia, hypomagnesemia –nausea and vomiting rarely
Pentostatin	–purine analogue; antimetabolite; inhibits adenosine deaminase **–nephrotoxicity (including acute renal failure)** –bone marrow suppression –neurotoxicity—lethargy, fatigue, seizures, coma –immunosuppression (lymphopenia) –nausea and vomiting –fever –anorexia –hepatotoxicity
Prednisone	–corticosteroid –leukocytosis –nausea and vomiting; indigestion –anorexia or increased appetite –CNS effects (depression, anxiety, euphoria, insomnia, psychosis, confusion) –fluid retention –hyperglycemia –osteoporosis –acne –adrenal insufficiency with prolonged use

Lung Cancer | Malignant Melanoma | Sarcoma | Supportive Care | Hematology Basics | Chemo-therapeutic Drug Toxicities | Drug Costs

Procarbazine	–alkylating agent; cell cycle independent **–bone marrow suppression—prolonged** –nausea and vomiting—severe; tolerance often develops with repeated dosing –mucocutaneous effects (mucositis, stomatitis, diarrhea) –rash, hives, photosensitivity –interstitial pneumonitis –CNS toxicity—seizures, lethargy, headache, ataxia –flu-like syndrome –azoospermia and amenorrhea almost universal
Rituximab	–monoclonal antibody to CD20 (B-cell surface antigen) –fever, chills, malaise –nausea, vomiting –flushing –bronchospasm, angioedema, urticaria –rhinitis –pain at disease sites –tumor lysis syndrome may occur in patients with high peripheral lymphocyte count
Sargramostim (GM-CSF)	–hematopoietic growth factor –nausea and vomiting –flushing –capillary leak syndrome –fevers and chills –headache –bone pain –myalgias, arthralgias –leukocytosis
Streptozocin	–alkylating agent; cell cycle independent **–nephrotoxicity—can be dose-limiting** –nausea and vomiting—may get progressively worse with continued administration –mucocutaneous effects (mucositis, stomatitis, diarrhea) –bone marrow suppression –irritant if extravasated (not vesicant) –delirium or depression –risk of secondary leukemias
Tamoxifen	–nonsteroidal antiestrogen –nausea and vomiting –bowel changes (diarrhea or constipation) –headache –peripheral edema –hot flashes –endometrial carcinoma –vaginal bleeding –venous thrombosis

Temozolomide	–alkylating agent; **–bone marrow suppression–delayed** –nausea and vomiting—mild to moderate –constipation –rash –headache –elevated transaminases
Teniposide (VM-26)	–topoisomerase II inhibitor **–bone marrow suppression** –nausea and vomiting –mucocutaneous effects (mucositis, stomatitis) –chemical phlebitis common –hypotension with rapid administration –hypersensitivity reactions –secondary leukemia
6-Thioguanine (6-TG)	–purine analogue antimetabolite; cell cycle dependent **–bone marrow suppression** –nausea and vomiting –mucocutaneous effects (mucositis, stomatitis) –rash –hepatotoxicity –hyperuricemia
Thiotepa	–alkylating agent; cell cycle independent **–bone marrow suppression** –nausea and vomiting–uncommon –mucocutaneous effects (mucositis, stomatitis)—uncommon –fever –angioedema –urticaria –secondary leukemia
Topotecan	–semisynthetic camptothecin; topoisomerase I inhibitor **–bone marrow suppression** –nausea and vomiting –mucocutaneous effects (mucositis, stomatitis) –constitutional symptoms—fatigue, anorexia, malaise –hematuria –renal insufficiency –hypertension –hepatotoxicity
Toremifene	– nonsteroidal antiestrogen –nausea and vomiting –bowel changes (diarrhea or constipation) –headache –peripheral edema –hot flashes –vaginal bleeding or discharge –venous thrombosis

Trastuzumab (Herceptin)	–humanized mouse monoclonal antibody directed against HER-2/*neu* receptor –fevers, chills, nausea, vomiting, headache during administration –cardiotoxicity (the FDA has not approved concurrent use with doxorubicin)

Tretinoin	–naturally occurring retinoid
	–retinoic acid syndrome
	  –fever
	  –chest pain
	  –hypoxia
	  –pulmonary infiltrates
	  –pleural/pericardial effusions
	–nausea and vomiting
	–mucocutaneous effects
	–arthralgias
	–headaches
	–increased triglycerides
	–xerostomia, exfoliation, chelitis

Trimetrexate	–antifolate antimetabolite **–bone marrow suppression** **–mucocutaneous effects (mucositis, stomatitis)** –nausea and vomiting –fever –maculopapular rash—usually self-limited –anorexia, malaise –above toxicities increased in patient with hypoalbuminemia (<3.5)
Vinblastine	–vinca alkaloid; inhibits tubulin polymerization; G2 phase specific **–bone marrow suppression** –vesicant if extravasated –nausea and vomiting –constipation (often secondary to neuropathy induced ileus) –neuropathy (jaw pain, peripheral neuropathy, autonomic neuropathy) –SIADH –tumor pain
Vincristine	–vinca alkaloid; inhibits tubulin polymerization; G2 phase specific **–neurotoxicity—peripheral neuropathy** –vesicant if extravasated –nausea and vomiting –bone marrow suppression—mild –transient transaminase elevation –constipation (often secondary to neuropathy induced ileus) ****–intrathecal injection is ALWAYS FATAL**

Vinorelbine	–vinca alkaloid; inhibits tubulin polymerization; G2 phase specific **–bone marrow suppression** –vesicant if extravasated –neurotoxicity –nausea and vomiting –acute reaction during administration—wheezing, chest pain, dyspnea –can be prevented on future administration with corticosteroids

Chapter 16
Hematology/Oncology Drug Costs

Hematology/Oncology Drug Costs

Below is a listing of commonly used drugs in the practice of Hematology and Oncology and their costs. This is not meant to be all-inclusive; it is meant to be a guide to the costs of the various drugs used in this field. If more than one dosage formulation is available for a specific agent, only the 1 or 2 most common forms are listed. In addition, even though common brand names are listed for recognition purposes, the cost reflects that of the lowest-priced generic (if one is available). Prices are those as of 11/99.

Oral Agents

AGENT NAMES	STRENGTH	FORM	COST ($)	COMMON BRAND
Altretamine	50 mg	capsule	6.62	Hexalen
Aminoglutethimide	250 mg	tablet	1.35	Cytadren
Anagrelide	0.5 mg	tablet	4.72	Agrelin
Anastrazole	1 mg	tablet	6.48	Arimidex
Bicalutamide	50 mg	tablet	11.53	Casodex
Busulfan	2 mg	tablet	1.82	Myleran
Capecitabine	500 mg	tablet	6.80	Xeloda
Chlorambucil	2 mg	tablet	1.58	Leukeran
Cyclophosphamide	50 mg	tablet	3.93	Cytoxan
Danazol	200 mg	tablet	2.50	Danacrine
Dexamethasone	2 mg	tablet	0.55	Decadron
	4 mg	tablet	0.37	
Dolasetron	100 mg	tablet	68.64	Anzemet
Estramustine	140 mg	capsule	3.83	Emcyt
Etoposide	50 mg	capsule	46.43	VePesid
Fluoxymesterone	5 mg	tablet	1.69	Halotestin
Flutamide	125 mg	capsule	2.02	Eulexin
Granisetron	1 mg	tablet	47.05	Kytril
Hydroxyurea	500 mg	capsule	1.03	Hydrea
Leucovorin	5 mg	tablet	2.35	Wellcovorin
Levamisole	50 mg	tablet	6.36	Ergamisol
Lomustine (CCNU)	100 mg	capsule	31.76	CeeNU
Medroxyprogesterone	10 mg	tablet	0.20	Provera
Megestrol	40 mg	tablet	0.85	Megace
	40 mg/ml	240 cc bottle	139.20	
Melphalan	2 mg	tablet	2.18	Alkeran
Mercaptopurine (6-MP)	50 mg	tablet	3.00	Purinethol
Methotrexate	2.5 mg	tablet	1.66	
Mitotane (o,p'DDD)	500 mg	tablet	2.69	Lysodren
Nilutamide	50 mg	tablet	2.81	Nilandron
Ondansetron	8 mg	tablet	26.47	Zofran
	24 mg	tablet	79.42	
Procarbazine	50 mg	capsule	0.69	Matulane
Tamoxifen	20 mg	tablet	3.53	Nolvadex

Continued

Oral Agents

AGENT NAMES	STRENGTH	FORM	COST	COMMON BRAND
Temozolomide	100 mg	capsule	120.00	Temodar
Thalidomide	50 mg	capsule	7.84	Thalomid
Thioguanine (6-TG)	40 mg	tablet	4.04	
Toremifene	60 mg	tablet	2.85	Fareston
Tretinoin (ATRA)	10 mg	capsule	11.88	Vesanoid

Injectable Agents

AGENT NAMES	AMOUNT IN VIAL	COST PER VIAL	COMMON BRAND
Aldesleukin (IL-2)	22 million IU	599.75	Proleukin
Amifostine	500 mg	1106.25	Ethyol
Antithymocyte globulin	25 mg	265.00	Thymoglobulin
Asparaginase	10,000 IU	60.43	Elspar
PEG-Asparaginase	3,750 IU	1391.20	Oncaspar
Bleomycin	15 unit	292.42	Blenoxane
Carboplatin	450 mg	899.42	Paraplatin
Carmustine (BCNU)	100 mg	104.36	BiCNU
Cisplatin	100 mg	454.90	Platinol
Cladribine	10 mg	562.80	Leustatin
Cyclophosphamide	1000 mg	49.36	Cytoxan
Cytarabine	500 mg	21.02	Cytosar-U
	2000 mg	98.90	
Dacarbazine (DTIC)	200 mg	23.14	DTIC-dome
Dactinomycin	0.5 mg	13.40	Cosmegen
Daunorubicin	20 mg	162.79	Cerubidine
Liposomal daunorubicin	50 mg	268.75	DaunoXome
Denileukin diftitox	300 mcg	992.50	Ontak
Dexamethasone	20 mg/ml	4.98	Decadron
Dexrazoxane	500 mg	296.30	Zinecard
Docetaxel	80 mg	1137.43	Taxotere
Dolasetron	100 mg	155.85	Anzemet
Doxorubicin	50 mg	225.40	Adriamycin
	100 mg	378.52	
Liposomal doxorubicin	20 mg	656.25	Doxil
Enoxaparin	30 mg	56.00	Lovenox
Epirubicin	50 mg	656.25	Ellence
Erythropoietin	40,000 units	480.00	Procrit
Etoposide	100 mg	44.00	VePesid
Etoposide phosphate	100 mg	119.19	EtopoPhos
Filgrastim	300 mcg	172.30	Neupogen
Fludarabine	50 mg	242.25	Fludara
Fluorouracil	1000 mg	3.00	Efudex
Gemcitabine	1000 mg	465.59	Gemzar
Goserelin	3.6 mg	469.99	Zoladex
	10.8 mg	1409.98	
Granisetron	1 mg	195.20	Kytril
Idarubicin	20 mg	1437.41	Idamycin
Ifosfamide	3000 mg	428.69	Ifex
Interferon alfa-2a	18 million IU	209.58	Roferon-A
Interferon alfa-2b	18 million IU	218.04	Intron-A
Irinotecan	100 mg	620.05	Camptosar
Lepirudin	50 mg	126.00	Refludan
Leucovorin	50 mg	56.25	Wellcovorin
	350 mg	85.75	

Continued

Lung Cancer
Malignant Melanoma
Sarcoma
Supportive Care
Hematology Basics
Chemo-therapeutic Drug Toxicities
Drug Costs

Injectable Agents

AGENT NAMES	AMOUNT IN VIAL	COST PER VIAL	COMMON BRAND
Leuprolide	7.5 mg	623.79	Lupron
	22.5 mg	1783.95	
Mechlorethamine	10 mg	11.59	Mustargen
Medroxyprogesterone	150 mg	48.10	Depo-Provera
Melphalan	50 mg	367.31	Alkeran IV
Mesna	2000 mg	368.80	Mesnex
Methotrexate	50 mg	4.36	
	250 mg	21.80	
Mitomycin C	20 mg	434.80	Mutamycin
Mitoxantrone	25 mg	1173.75	Novantrone
Octreotide	0.5 mg	56.80	Sandostatin
Octreotide long acting	20 mg	1368.75	Sandostatin LAR Depot
Ondansetron	40 mg	256.40	Zofran
Oprelvekin	5 mg	248.75	Neumega
Paclitaxel	300 mg	1826.25	Taxol
Pamidronate	90 mg	678.31	Aredia
Pentostatin	10 mg	1440.00	Nipent
Rh_o (D) Immune Globulin	300 mcg(1500 IU)	306.00	WinRho
Rituximab	500 mg	2212.08	Rituxan
Sargramostim (GM-CSF)	250 mcg	134.85	Leukine
Streptozocin	1000 mg	114.65	Zanosar
Teniposide	100 mg	394.68	Vumon
Thiotepa	15 mg	105.58	Thioplex
Topotecan	4 mg	603.95	Hycamtin
Trastuzumab	440 mg	2262.50	Herceptin
Trimetrexate	25 mg	73.50	Neutrexin
Vinblastine	10 mg	21.25	Velban
Vincristine	2 mg	29.24	Oncovin
Vinorelbine	50 mg	381.45	Navelbine

Appendix—Miscellaneous Formulas

Calvert Formula
–used for AUC dosing of Carboplatin

$$\frac{(140\text{-age}) \times \text{weight in kg} \times (0.85 \text{ in females, } 1.0 \text{ in males})}{72 \times \text{serum creatinine}} = \text{estimated creatinine clearance}$$

estimated CrCl + 25 = GFR

GFR X target AUC = Carboplatin dose

Performance Status
Karnofsky

100	normal
90	minor signs/symptoms of disease
80	some signs/symptoms of disease
70	cares for self; unable to carry on normal activity or actively work
60	requires occasional assistance
50	requires considerable assistance
40	disabled; requires special care
30	severely disabled; hospitalization is indicated; death is not imminent
20	very sick; hospitalization necessary
10	moribund
0	dead

ECOG

0	fully active (90-100)
1	restricted to light activities (70-80)
2	capable of self-care (50-60)
3	limited self-care; confined to bed or chair >50% of waking hours (30-40)
4	completely disabled (10-20)
5	dead (0)

Index

G

H

I

K

L

M

N

O

P

R